W9-ACT-567

UNIVERSITY
OF ROCHESTER
LIBRARY

Presented by
Dr. Neil E. Wahlberg
Class of 1972 Acquisition Fund

LIE GROUPS: HISTORY, FRONTIERS AND APPLICATIONS

SERIES B: SYSTEMS INFORMATION AND CONTROL

VOLUME I

GEOMETRY AND IDENTIFICATION:

PROCEEDINGS OF APSM WORKSHOP ON SYSTEM GEOMETRY, SYSTEM IDENTIFICATION, AND PARAMETER ESTIMATION

MAY 18-22, 1981

EDITORS

PETER E. CAINES

ROBERT HERMANN

MATH SCI PRESS

1983

Copyright © 1983 by Robert Hermann
All rights reserved

ISBN: 0-915692-33-3

Library of Congress Cataloging in Publication Data

APSM Workshop on System Geometry, System Identification, and Parameter Estimation (1981 : Northeastern University)
Geometry and identification.

(Lie groups. Series B, Systems information and control ; v. 1)
Includes bibliographies.
1. System identification--Congresses. 2. Parameter estimation--Congresses. 3. Geometry--Congresses. I. Caines, Peter E., 1945- . II. Hermann, Robert. III. Association for Physical and Systems Mathematics (U.S.) IV. Title. V. Series.
QA402.A67 1981 003 83-769
ISBN 0-915692-33-3

MATH SCI PRESS

53 JORDAN ROAD

BROOKLINE, MASSACHUSETTS 02146

Printed in the United States of America

TABLE OF CONTENTS

PREFACE

The APSM Workshop on System Geometry, System Identification and Parameter Estimation was held at Henderson House, the Northeastern University conference center in Weston, Massachusetts, in May 1981. These surroundings provided a pleasant and stimulating environment for the meeting, and we take this opportunity to thank Professor Gustav Schacter of Northeastern University and the Conference secretaries, Lorraine Luccesi, Joyce Martin and Karin Young. The next workshop, titled "Nonlinear Problems in Control and Fluid Dynamics", is planned for June 1983 at the University of California, Berkeley.

As the title indicates, this was an interdisciplinary meeting between research workers in the system theory area of theoretical engineering. The motivation was the following: Over recent years there has been speculation that the increasingly sophisticated results and ideas of geometric system theory would be useful in the theory of system identification and adaptive control. Conversely, it has been suspected that new geometric problems will arise as the qualitative features of system identification and adaptive control algorithms become better known. The purpose of this workshop was to bring together research workers in these fields in the belief that a fruitful interaction between these areas could now take place. We were enabled to do this by the support of grants from the National Science Foundation and the Ames Research Center of the National Aeronautics and Space Administration.

The conference was organized so as to interleave talks by members of each research culture. Morning talks consisted of key-note and review lectures, while afternoon talks were of a more specialized nature.

The papers in this volume are based upon the talks given at the meeting and we give below a list of the authors, the titles of their articles and a brief outline of their contents.

P.E. Caines: Modelling and Maximum Likelihood Estimation of Gaussian ARMAX and State Space Systems.

Three basic topics are presented in a tutorial fashion. First, the construction of parameterized models of systems driven by and generating Gaussian processes, where these systems are, in addition, driven by an observed ("exogenous") deterministic or random process. The manifold structure of the parameter space for minimal realizations enters here and a suitable definition is formulated of identifiability for systems with both stochastic and deterministic behavior. Second, the construction of likelihood functions on the observed processes is described together with an example, and, third, a principal theorem concerning the asymptotic (large sample) properties of maximum likelihood estimates of the system parameters is presented.

D.F. Delchamps: Geometric Questions in System Identification.

Many system identification algorithms operate by minimizing some form of objective function V computed in terms of the observed input-output data on the given system together with the parameter $\sigma \in \Sigma$ of each candidate system. Often, such algorithms may be modelled as the flow induced on the manifold of system parameters Σ by the gradient vector field - grad V taken with respect to some Riemannian metric on Σ. In this paper it is shown that if either the number of system inputs m or outputs p exceeds 1, there does not exist a function $V: \Sigma \to \mathbb{R}$ with a single critical point which is locally and globally attracting for the vector field - grad V. Further, results in this paper give information on the homology groups of Σ when $\min(m,p) > 1$ and on the number of critical points that an objective function on Σ must necessarily possess.

B.W. Dickinson: Structure of Stationary Finite Observation Records of Discrete-Time Stochastic Linear Systems.

The existence of nontrivial sufficient statistics is studied for the parameters of ARMA systems driven by white noise given observations on the system output. It is shown that only autoregressive processes have sufficient statistics whose dimension is less than the number of observations. This fact is linked to the theory of stochastic realization.

M. Diestler: The Structure of ARMA Systems in Relation to Estimation.

The parameterization of ARMA systems by matrix fraction descriptions is studied and the manifold structure of all systems of given order is described. In particular, the specification of the echelon canonical form via the system dynamical indices is presented in detail. Next, the topological and geometric properties of the space of system parameterizations and its boundary points are discussed for the so-called structural identifiability, echelon canonical form and overlapping description representations, respectively. The relation with the topology on the corresponding sets of transfer functions is considered.

Finally, the use of these various representations in system order and system parameter estimation is analysed.

T.E. Duncan: Some Geometric Methods for Stochastic Integration in Manifolds.

This paper lays the foundations for the analysis of stochastic systems in manifolds. A comparison is made of the various extant definitions of stochastic integrals formed by pairing one-forms with a semi-martingale in a Riemannian manifold. By using the canonical one-form of the linear connection, an approach is introduced which corresponds naturally with the methods used in Euclidean spaces and with the study of geodesics in manifolds. Among other results, a local coordinate description of such a Brownian semi-martingale is given and an appropriate generalization of the Girsanov theorem is presented.

M. Gevers and V. Wertz: Overlapping Parameterizations for the Representations of Multivariate Stationary Time Series.

When identifying a state space of ARMA model for a multivariate stationary stochastic process, the first problem is to find the order (equivalently, state space dimension) of the model and define a uniquely identifiable parameterization. For a given order, several overlapping parameterizations, all involving the same number of parameters, can usually be associated to the same process. An asymptotic result is presented showing that all overlapping parameterizations give the same value to the determinant of the Fisher information matrix and so, for many identification schemes, to the determinant of the asymptotic parameter estimation error covariance matrix. For finite data some structures may be preferable to others. Two heuristic estimation methods are introduced and discussed along with some simulation results.

R. Hermann and C.F. Martin: Lie Theory of Transformation Groups and the Parameterization and Identification of Linear Systems.

Lie's general concept of the action of a transformation group on the set of solutions of a (family of) differential equations is outlined. The manifold structure of the orbits of general transformation groups are examined and an order relation defined via the closure of the orbit types in the finite dimensional situation. Detailed information is obtained for positive Riemannian manifolds. This apparatus is then applied to finite dimensional time invariant linear systems and results on the dimension of the orbits, orbit spaces mode Gl(n) and isotropy groups reviewed.

O. Hijab: Bayesian Systems Identification in the Presence of Small Noise.

For continuous time parameter linear stochastic systems the likelihood function on the parameters, given observations over a finite interval, is derived along with the Bayesian distribution in the case where the parameters are distributed independently of all other processes. Both are computed for systems subject to a wide class of feedback control laws. The behavior of the Bayes estimator when the variances of the random initial state, system and observation noise are scaled by ε is examined. It is shown that as $\varepsilon \downarrow 0$, the Bayesian estimate converges to the maximum likelihood estimate.

S.E. Landsberg: Algebraic Geometry and the Business Cycle.

Models arising from the consideration of rational expectations in economic behavior are developed in this paper. The question of the identifiability of these models is then addressed and the role of algebraic geometry in model specification in the typical non-identifiable situation is discussed.

A.J.M. van Overbeck and L. Ljung: On-Line Structure Selection for Multivariable State Space Models.

An algorithm is described for the selection of a model structure for identifying state space models. The algorithm receives as "input" a given system described in a certain parameterization. This is then tested to see whether this parameterization is suitable (well conditioned) for identification purposes. If not, a better one is selected and the transformation of the system to the new representation is performed. The algorithm can be called whenever there is some indication that the model structure is ill conditioned and, furthermore, it may be interfaced with an off-line identification procedure. Tests on real and simulated data are described.

A. van der Schaft: Representations of Nonlinear Systems: Minimality, Observability and Controllability.

For electrical networks, and some physical and economic systems, it is not immediately clear which of the external variables are causes (inputs) and which are effects (outputs). For smooth nonlinear systems with external variables a definitive framework is provided so that it is not necessary, a priori, to distinguish between inputs and outputs; indeed, such a separation is only possible locally. For the representations introduced in this paper, a review is given of some results on minimality, observability and controllability.

Abstracts

H.M. Hastings and R. Pekelney: Stochastic Information Processing in Biology.

J.C. Willems: Realization and Representation of General Dynamical Systems.

Participants

K. Astrom
R. Bach
J. Baillieul
B. Balaban
M. Babas
R. Brockett
P. Caines
B. Chan
R. Cline
M. Deistler
D. Delchamps
B. Dickinson
D. Donoho
A. Doonovskoy
T. Duncan
W. Dunsmuir
R. Fogel
M. Gevers
K. Glover
R. Grossman
H. Hastings
R. Hermann
O. Hijab
G. Hyman
S. Kahne
C. Keng
P. Kumar
P. Krishnaprasad
S. Landsburg
W. Larrimore
S. Marcus
C. Martin
G. Mealy
R. Mehra
T. Overbeck
M. Pavon
R. Poetscher
R. van der Schaft
V. Solo
R. Stevens
G. Verghese
J. Willems
W. Wong

	9:00 10:30	10:45 12:15	LUNCH	2:00	3:00 3:45	4:00	4:45 5:30
MONDAY	Peter E. Caines	C. Byrnes		T. van Overbeek	O. Hibja	P. Krumar	B. Dickinson
TUESDAY	R.W. Brockett	K.J. Aström	S. Kahne	M. Pavon	K. Glover	M. Balas	W. Larrimore
WEDNESDAY	M. Deistler	W. Dunsmuir	H. Hastings	P. Krishnaprasad	R.W. Brockett	S. Landsberg	Discussion
THURSDAY	M. Gevers	J. Willems	R. Mehra	D. Delchamps		A. Van der Schaft	W. Wong
FRIDAY	R. Hermann C. Martin	T. Duncan P. Caines					

PROGRAM

WORKSHOP ON SYSTEM GEOMETRY, SYSTEM IDENTIFICATION AND PARAMETER ESTIMATION

MODELLING AND MAXIMUM LIKELIHOOD ESTIMATION OF GAUSSIAN ARMAX AND STATE SPACE SYSTEMS

P. E. Caines

This paper presents three topics in a tutorial fashion: first, the construction of parameterized models of systems driven by and generating Gaussian processes; second, the construction of likelihood functions on the observed input-output processes of such systems and, third, a principle (consistency and asymptotic normality) theorem concerning the large sample properties of the maximum likelihood estimates of the parameters of such systems.

The first consistency proof for maximum likelihood estimators in the independent identically distributed observations case was presented in Wald's classic paper [Wald, 1949], while current proof techniques are rigorized versions of the earlier method of Kendall and Stuart [Kendall and Stuart, 1948]. The analysis of maximum likelihood estimators in the AR (autoregressive) process case was carried out in 1943 by Mann and Wald [Mann and Wald, 1943] and for the MA (moving average) case by Whittle [Whittle, 1952], Durbin [Durbin, 1959] and Walker [Walker, 1961]. (See also the text by Anderson [Anderson, 1971]). The pioneering work of Aström and Bohlin [Aström and Bohlin, 1965] extended the analysis of Kendall and Stuart to the ARMAX (autoregressive moving-average system with exogeneous, i.e. observed, inputs) case; it introduced the techniques of nonlinear programming via digital computer for finding such maximum likelihood estimates in practice. This idea was also suggested by Mayne [Mayne, 1966]. Later, the well-known book by Box and Jenkins [Box and Jenkins, 1970] made an integrated set of parameter and system order estimation techniques for ARMA systems available to a wide audience.

Stemming from the advances which have just been sketched out, there was a proliferation of identification techniques for linear stochastic systems during the nineteen seventies; this work was carried out within statistics and time series analysis, engineering and econometrics. As far as the theory of maximum likelihood estimators for ARMA systems is concerned, it is believed that the first complete consistency results were given by J. Rissanen and the author in Caines and Rissanen [1974], Rissanen and Caines [1979]. An important parallel line of work, containing more extensive results, is that of Dunsmuir, Hannan and Deistler [Dunsmuir and Hannan, 1976; Deistler, Dunsmuir and Hannan, 1978]. (See also Hannan [1979]; Dunsmuir [1979]; Hannan, Dunsmuir, Deistler [1980].)

A feature of the analysis of the work by Rissanen and the author is that it uses linear estimation theory to give a convenient representation of the likelihood function for ARMAX processes. This method was apparently first used by Schweppe [Schweppe, 1973] and Mehra [Mehra, 1970, 1971]. This technique facilitates the analysis in the cases under consideration in this paper which are, in a precise sense, asymptotically stationary; furthermore, it permits a treatment of maximum likelihood estimators in the nonstationary case which would otherwise seem to be difficult (see Goodrich and Caines [1979b]).

This paper deals solely with the maximum likelihood method for Gaussian systems. The natural generalization of this approach is the prediction error identification method (see eg., Ljung [1976]; Caines [1976]; Ljung and Caines [1979]). With this approach all of the main hypotheses of the maximum likelihood method--linearity, Gaussian distributions, stationary input processes, exact modelling, etc.--are relaxed and appropriate versions of the consistency and asymptrophic normality results are obtained.

1. GAUSSIAN ARMAX AND STATE SPACE MODELS. We begin by considering a class $\mathcal{Y}$ of p component stochastic processes, defined on the probability space $(\Omega,\mathcal{B},P)$, where $y \in \mathcal{Y}$ is the sum $\zeta_k + \xi_k$, $k \in \mathbb{Z}$, of a process ζ, generated nonanticipatively by observed inputs in the class $\mathcal{U}$, and a process ξ in the class Ξ, generated by unobserved inputs, i.e.

$$y_k = \zeta_k + \xi_k, \qquad k \in \mathbb{Z} \quad , \tag{1.1}$$

where $\zeta_k = \zeta_k(u^k_{-\infty})$, $u_k = u_k(\omega)$, for all $k \in \mathbb{Z}$.

Our treatment of the identification problem, for the system mapping u to y, will only include linear systems that have finite dimensional realizations because these map Gaussian processes to Gaussian processes, and, in a sense to be made precise, they have finite dimensional parameterizations. Further, we are interested in longitudinal asymptotic properties of the estimators; this stands in contrast to the cross-sectional asymptotic properties examined in Goodrich and Caines [1979b]. These considerations lead us to restrict ourselves to asymptotically stable linear time invariant systems, since they are definitely capable of some form of asymptotically stationary behaviour, whether driven by deterministic or stochastic processes. Once this restriction is accepted, there is the prospect of an asymptotic analysis of the behaviour of various estimators via properties of the strong law and ergodic type.

The process ζ is taken to be the output process of an asymptotically stable finite dimensional linear time invariant system described by the $(p \times m)$ transfer function matrix $Z(z) = \sum_{i=0}^{\infty} Z_i z^i$. The m component input process u to Z is directly observed, defined on $(\Omega,\mathcal{B},P)$, and is assumed to satisfy one of the conditions:

INP 1A: u is deterministic (i.e., $\{\phi,\Omega\}$ measurable) and bounded.
INP 1B: u is a full rank strictly stationary ergodic process.

Here, and elsewhere in this paper, we shall take sequences of constants, as in INP 1A above, to be $\{\phi,\Omega\}$ measurable functions on Ω. This is done to give a certain formal completeness to the presentation.

Subject to INP 1 the sequence of sums $\zeta \underline{\Delta} \{\xi_k \underline{\Delta} \sum_{i=0}^{\infty} Z_i u_{k-i}; k \in Z\}$ exists as either (A) a bounded deterministic sequence, or (B) a wide sense stationary process.

Turning to the class of processes Ξ, which are generated by the unobserved inputs, we again look for the properties of finite dimensional parameterizability (for exact modelling) and for some form of stationary (for longitudinal asymptotic results).

A further modelling restriction to be placed on processes in Ξ is that they have no linearly deterministic part, that is to say, ξ_k cannot be decomposed as $\xi_k^s + \xi_k^d$ with $\xi_k^d = (\xi_k^d | H_{k-1}^{\xi})$ a.s. In the case of a wide sense stationary process, for instance, this means that $\underline{H}_{-\infty}^{\xi} = \phi$, that is to say the intersection of the spaces $\overline{\text{Span}}\,\{\ldots\xi_{k-1},\xi_k\}$ is empty, where overbar denotes closure.

This assumption is made purely for the sake of simplicity. Models involving nonstationary deterministic parts, such as trends (e.g., $\xi_k = \sum_{i=0}^{m} \alpha_i(\omega)k^i$), or stationary deterministic parts, such as periodicities (e.g., $\xi_k = A(\omega)\sin(a(\omega) + \frac{2\pi k}{N})$), are most important in applications such as econometrics and meterology. Furthermore, system theory based techniques are having an impact on parameter estimation for such processes (see e.g., Gersch [1981], Chan, Goodwin, Sin [1982]). However, we shall exclude consideration of such topics in order to keep our exposition to a manageable scale.

The same holds for the mean value process of the observed process y: In order to keep the system modelling theory of this paper simple--avoiding complicated "extensions" of the basic idea--we shall consider only those nonzero biases that can be represented by the signal ζ generated from u via Z.

A penultimate, technical, step in this discussion is to observe that although any p component process ξ under consideration is assumed to have no deterministic part, it still need not be of full rank. (Even in the wide stationary case this occurs if and only if $\int_0^{2\pi} \log \det F_{\xi}'(\theta)d\theta > -\infty$ by the Kolmogorov-Wiener-Masani Theorem.) Now if nonzero linear functionals of ξ are linearly predictable, i.e. if ξ is not full rank, then the same is true for y, since ζ is exactly predictable given the impulse response of the system Z and the input u. In this case, it is a perfectly reasonable simplification of the identification problem just to take a maximal subset of the components of y for which there are no such nonzero linear functionals that are exactly predictable. If we assume this has been done we are free to assume that ξ is of full rank.

Obviously one of the most important types of processes that may satisfy the constraints on ξ is the class of wide sense stationary processes. Now, by one of the principal results of stochastic realization theory, ξ is a p component full rank zero mean wide sense stationary process with a rational spectral density matrix Φ, for which $\det \Phi$ does not vanish on $T = \{z : |z| = 1\}$, if and only if ξ has the a.s. representation

$$\xi_k = \sum_{i=0}^{\infty} W_i w_{k-i} , \qquad W_0 = I, \quad k \in \mathbb{Z} \tag{1.2}$$

where $W(z)$ is a $p \times p$ asymptotically stable, inverse asymptotically stable matrix of rational functions and w is a p component zero mean wide sense stationary orthogonal process with a positive definite covariance matrix Σ.

It is convenient at this point to isolate the following assumptions on the process w:

<u>INP 2</u>: The process w defined on $(\Omega, \mathcal{B}, P)$ is a zero mean wide sense stationary orthogonal process with $Ew_k w_j^T = \Sigma\delta_{kj}$, for all $k,j \in \mathbb{Z}$, with $\Sigma \in \mathcal{P}$, where $\mathcal{P}$ denotes the set of $p \times p$ positive definite matrices.

Now the system W, when it acts on a process w satisfying INP2, is able to generate wide sense stationary processes satisfying the hypotheses we have declared to be desirable. For this reason, we settle on systems with the properties of W as the characterization of the part of the system generating ξ. It should be noticed, however, that we shall use this assumption from here on *without* assuming ξ to be wide sense stationary. (Precisely: W will act on w for $k \in \mathbb{Z}_+$, but the system will not in general be given the initial state covariance corresponding to a wide sense stationary state and output process.)

Following this discussion, which has lead us from the properties of the process class $\mathcal{Y}$ to an admissable set of system models for generating certain members of $\mathcal{Y}$, we concentrate on the description of systems themselves.

Since ζ and ξ in (1.1) have the form Zu and Ww respectively,

$$y(z) = [Z(z)W(z)] \begin{pmatrix} u(z) \\ w(z) \end{pmatrix} , \tag{1.3}$$

with the matrix $[Z(z)W(z)]$ being a $(p \times (m+p))$ matrix of rational functions of z such that all entries have asymptotically stable poles, $\det W(z)$ has asymptotically stable zeros and $W(0) = I$.

At this point we restrict attention to transfer functions $[Z(z)W(z)]$ which are proper i.e. $\lim_{z\to\infty} [(Z(z)W(z)] < \infty$. This is done purely to simplify the discussion.

In this case, we know from standard realization theory (see e.g. Desoer [1970], Heymann [1975]), that

ARMAX 1

$[Z(z)W(z)]$ has the left matrix fraction description

$$[Z(z)W(z)] = A^{-1}(z)[B(z)C(z)] \tag{1.4}$$

where A,B,C are respectively $(p\times p)$, $(p\times m)$ and $(p\times p)$ polynomial matrices such that $A^{-1}(z)$ exists almost everywhere, $\det A(z)$ and $\det C(z)$ have asymptotically stable zeros, $A(z)$ and $[B(z)]$ have no common left matrix divisors other than unimodular matrices (i.e. $A(z)$ and $[B(z)C(z)]$ are left coprime), or, equivalently, $\deg \det A(z)$ in (1.4) has the smallest (integer) value δ amongst all such representations, the value of δ being the Smith-McMillan degree of $[Z(z)W(z)]$. Furthermore, every row of $[B(z)C(z)]$ has degree less than the corresponding row of $A(z)$.

Finally, $A^{-1}(0)\, C(0) = I$.

Equivalently,

SSX 1

$[Z(z)W(z)]$ has the representation

$$[Z(z)W(z)] = H(Iz^{-1} - F)^{-1}[G^uG^w] + [D^uI] \quad , \tag{1.5}$$

where $(H,F,[G^uG^w],[D^uI])$ are $(p\times\delta)$, $(\delta\times\delta)$, $\delta\times(m+p)$ real matrices respectively, F is asymptotically stable, $\det(H(Iz^{-1} - F)^{-1}G^w + I)$ has asymptotically stable zeros, (H,F) is observalbe and $(F,[G^uG^w])$ is controllable, or, equivalently, the state space dimension δ in (1.5) is the smallest amongst all such state space realizations, the value of δ being the Smith-McMillan degree of $[Z(z)W(z)]$.

The initials SSX here stand for "state space system with exogeneous inputs."

If we do not assume $[Z(z)W(z)]$ is proper, then we simply lose the row degree property of ARMAX 1 and must add pure delay terms to the representation (1.5).

We now specify our admissable *set* of ARMAX and state space systems by considering the *set* of linear systems having the properties ARMAX 1 or SSX 1 above. To be specific we adopt

DEFINITION 1.1. *For given* $p,m,\delta \in \mathbb{Z}_+$, $S(p,m,\delta)$ *denotes the set of linear time invariant finite dimensional systems each of whose members is described by a* $p\times(m+p)$ *transfer function* $[Z(z)W(z)]$ *such that there exists*

(1) *A Matrix Fraction Description* $(A(z),[B(z)C(z)])$ *satisfying ARMAX 1, or, equivalently,*

(2) *A State Space Description* $(H,F,[G^uG^w],[D^uD^w])$ *satisfying SSX 1.*

When $m=0$ this will indicate that we are dealing with the case where u is absent.

Now a parameterizing space for $S(p,m,\delta)$, without the asymptotic stability constraints appearing in ARMAX 1 or SSX 1, is provided by an (analytic) manifold M^ν of dimension ν. This is the fundamental result of Clark [J. M. C. Clark, 1976]. (See also Rissanen and Ljung [1976], Hazewinkel and Kalman [1975], Hazewinkel [1976], Byrnes [1976], and Hannan [1979].)

By imposing the asymptotic stability constraints of ARMAX 1 or SSX 1 we restrict ourselves to a certain submanifold N of M^ν. We now state

DEFINITION 1.2. $\Psi(p,m,\delta)$ *shall denote the parameter space of* $S(p,m,\delta)$ *consisting of the submanifold* $N \subset M^\nu$.

$\Psi(p,m,\delta)$ shall be written Ψ whenever the context leaves no ambiguity concerning p,m and δ.

Finally we take the $p \times p$ covariance matrix Σ of the orthogonal process w and let it vary over the set P of $p \times p$ symmetric positive definite matrices. For given p,m,δ this results in $\Theta \underline{\Delta} \Psi \times P$ being the parameter space for the systems appearing in (1.3), satisfying ARMAX 1 or SSX 1, and being driven by an orthogonal process with covariance $\Sigma > 0$. (The one-to-one relationship between elements of $S(p,m,\delta) \times P$ and elements of Θ is shown in Fig. 1.1.) In an obvious notation we write the system equations as

$$\text{ARMAX:} \begin{cases} A_\psi(z)y(z) = B_\psi(z)u(z) + C_\psi(z)w(z)\ , \quad \psi \in \Psi \quad \forall k \in \mathbb{Z}\ , & (1.6a(i)) \\ Ew_k w_j^T = \Sigma\delta_{kj}\ , \quad \forall k,j \in \mathbb{Z}\ , \quad \Sigma \in P & (1.6a(ii)) \end{cases}$$

and

$$\text{SSX:} \begin{cases} x_{k+1} = F_\psi x_k + G_\psi^u u_k + G_\psi^w w_k\ , & (1.6b(i)) \\ y_k = H_\psi x_k + D_\psi^u u_k + Iw_k\ , \quad \psi \in \Psi \quad \forall k \in \mathbb{Z}\ , & (1.6b(ii)) \\ Ew_k w_j^T = \Sigma\delta_{kj}\ , \quad \forall k,j \in \mathbb{Z}\ , \quad \Sigma \in P\ . & (1.6b(iii)) \end{cases}$$

$$\left\{\begin{array}{l} A(z),[B(z)C(z)]) \\ \text{satisfying ARMAX 1,} \\ \text{plus } \Sigma > 0 \end{array}\right\} \overset{1:1}{\longleftrightarrow} \left\{\begin{array}{c} \Theta \\ \theta \in \Theta \end{array}\right\} \overset{1:1}{\longleftrightarrow} \left\{\begin{array}{l} (H,F,[G^u G^w],[D^u I]) \\ \text{satisfying SSX 1,} \\ \text{plus } \Sigma > 0 \end{array}\right\}$$

Figure 1.1

(1.6a) and (1.6b) constitute our fundamental set of linear stochastic systems. It is the parameters $\theta \in \Theta$ for members of this set which we wish to estimate from observations on (y,u) using the most effective means we can find.

The notion of *identifiability* for a system whose inputs may be deterministic or stochastic and observed or unobserved must be formulated with some care. We shall state

DEFINITION 1.2. *The system parameterized by* Θ *in* (1.6) *is said to be identifiable if there exists a (*u *dependent) one-to-one mapping between*

(i) *The system data:* (a) *the impulse response (Markov matrices) of the system* Z,

(b) *the covariance data of the process* $y - Zu$,

where (y,u) *are related by* (1.6), *the system* $[Z\ W]$ *lies in* $S(p,m,\delta)$ *and* (u,w) *satisfies the input hypotheses* A *or* B*; and*

(ii) *the system parameter space* Θ.

This definition satisfies the intuitive requirement that a parameterized system is identifiable if there is a one-to-one relationship between the system data and the system parameter set (see e.g., Goodrich and Caines [1979a] and the references therein). Notice that this definition involves the structure of the parameterized system, the parameterization and certain properties (in this case second order properties) of the stochastic processes; inferential methods and the properties of estimators do not, and should not, appear in a definition of identifiability. The identifiability property holds in the case under consideration because starting from (i), we may pass, by virtue of our specification of $S(p,m,\delta)$, in a one-to-one manner to the matrices of rational functions $[Z(z)W(z)]$, corresponding to systems in $S(p,m,\delta)$, together with the covariance matrix $\Sigma \in P$. Then, by virtue of our specification of Θ, we may pass in a one-to-one manner from $[Z(z)W(z)]$, Σ to an element of Θ. The nature of this identifiability relationship, factored through the set $S(p,m,\delta) \times P$, is illustrated in Figure 1.2.

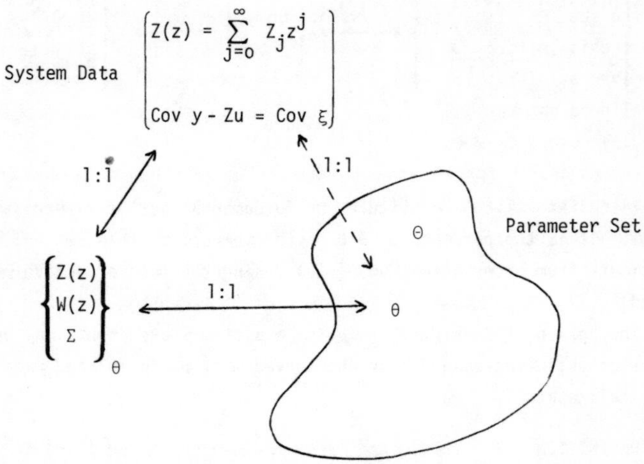

Figure 1.2

We take the topology on Θ to be the product topology inherited from the constitute space Ψ, endowed with its manifold topology, and the space $\mathcal{P}$, endowed with the topology of Euclidean space.

Given infinite data, generated by a system in $\mathcal{S}(p,m,\delta)$, the algebraic identifiability relation, together with the assumptions on the joint process (u,w), will permit us to pass, in the limit, from the set of observations back to the system generating the data. An alternative way to state this is as follows: given infinite observations on a process y of the form $y = Zu + \xi$, with Z having a rational asymptotically stable transfer function, u satisfying INP 1, and ξ satisfying the given set of conditions, then it is possible to pass to the single parameterized model in $\mathcal{S}(p,m,\delta)$ that generated the observed path. This is an "almost sure" statement and, in any case, we have not proved it yet. That comes in Section 3.

2. CONSTRUCTION OF THE LIKELIHOOD FUNCTION. Observations on the inputs and outputs of systems such as ARMAX and SSX ((1.6a & b) of the previous section) over the interval $[1,N]$ constitute a set of random vector variables $\{z_1,z_2,\ldots,z_N\} \triangleq \{(y_1,u_1),(y_2,u_2),\ldots,(y_N,u_N)\}$ which we may arrange into the vector variable

$$(z_1^N)^T = (y_N^T,u_N^T;y_{N-1}^T,u_{N-1}^T;\ldots;y_1^T,u_1^T)^T \in \mathbb{R}^{N(m+p)} \quad .$$

Let us assume for the moment that the random variable z_1^N is distributed over

$\mathbb{R}^{N(m+p)}$ with the parameterized density $f(z_1^N;\overset{\circ}{\theta})$, where $\overset{\circ}{\theta}$ takes a value in the set Θ. (The density assumption would rule out INP 1A as the input hypothesis in force.) Then the likelihood function of the observations z_1^N is given by $f(z_1^N;\cdot):\Theta \to \mathbb{R}^1$, and by the subsequent Definition 2.2, the maximum likelihood estimate $\hat{\theta}_N$ (of $\hat{\theta}$) is given by the maximum (when it exists) of $f(z_1^N;\theta)$ over Θ.

Let us assume for the moment that z is a full rank zero mean stationary Gaussian stochastic process with (infinite) process covariance matrix $0 < \Sigma^z = (\Sigma^z_{i,j} = \Sigma^z_{i=j};\ i,j \in \mathbb{Z})$, where $\Sigma^z_{i,j} = Ez_i,z_j^T$, for all $i,j \in \mathbb{Z}$. Then $f(\cdot;\cdot)$ is parameterized by $\Sigma^z \in \Theta$ when we identify Θ with the set of $Np \times Np$ positive definite matrices. Then we have

$$f(z_1^N;\theta) = (2\pi)^{-Nr/2}(\det \Sigma_N^z(\theta))^{1/2} \exp - \tfrac{1}{2}\{(z_1^N)^T[\Sigma_N^z(\theta)]^{-1}(z_1^N)\} \quad , \qquad \theta \in \Theta \tag{2.1}$$

where $r = m+p$ and where the subscript N of $\Sigma_N^z(\cdot)$ denotes the $Nr \times Nr$ submatrix on the diagonal of $\Sigma^z(\cdot)$ corresponding to z_1^N.

Returning to our main theme of the time domain construction of the exact likelihood function, let us assume the random variable $\{y_1^N\}$ defined on $(\Omega,\mathcal{B},P)$ has a density depending upon the deterministic quantities (u_1^N,θ) for all N. (We have in mind here the case of observations generated by input subject to INP A). Then using Bayes rule, or, equivalently, the definition of conditional densities, we obtain

$$f(y_1^N;u,\theta) = f(y_N|y_1^{N-1};u,\theta)f(y_1^{N-1};u,\theta) = \prod_{i=1}^{N} f(y_i|y_1^{i-1};u,\theta) \tag{2.2a}$$

where $f(\cdot;u,\theta)$ is used generically to denote the indicated conditional densities (which necessarily exist) and where the second equality follows by induction.

Alternatively, let (y_1^N,u_1^N) have a joint density depending upon the deterministic quantity. (Here we have in mind INP B). Further assume the conditional densities $f(u_k|u_1^{k-1},y_1^{k-1};\theta)$ are not functions of θ for each $k \in \mathbb{Z}_1$. Reasoning as before, we obtain

$$\begin{aligned} f(y_1^N,u_1^N;\theta) &= f(y_N,u_N|y_1^{N-1},u_1^{N-1};\theta)f(y_1^{N-1},u_1^{N-1};\theta) \\ &= f(y_N|y_1^{N-1},u_1^N;\theta)f(u_N|y_1^{N-1},u_1^{N-1};\theta)f(y_1^{N-1},u_1^{N-1};\theta) \\ &= \prod_{i=1}^{N} f(y_i|y_1^{i-1},u_1^i;\theta)f(u_i|y_1^{i-1},u_1^{i-1}) \quad . \end{aligned} \tag{2.2b}$$

We see that $f(y_1^N;u,\theta)$ in (2.2a) is the likelihood function of the data y_1^N. The function is parameterized by the deterministic quantities (u,θ)

where u is $\{\phi,\Omega\}$-measurable and θ is a ("purely" deterministic) element of Θ. Further, $f(y_1^N,u_1^N;\theta)$ in (2.2b) is the likelihood function of (y_1^N,u_1^N) where hence θ is the only deterministic quantity parameterizing the function.

Now let us write

$$\hat{y}_{i|i-1}(\theta) = \int_{\mathbb{R}^p} y_i \, f(y_i|y_1^{i-1};u,\theta)dy_i \in \mathbb{R}^p$$

(the conditional expectation of y_i given y_1^{i-1}), in case (u,θ) are deterministic, and let

$$y_{i|i-1}(\theta) = \int_{\mathbb{R}^p} y_i \, f(y_i|y_1^{i-1},u_1^i;\theta)dy_i \in \mathbb{R}^p \quad ,$$

(the conditional expectation of y_i given (y_i^{i-1},u_1^i)), in case (y,u) are jointly distributed and θ is deterministic.

Making the change of variables

$$y_i \to \hat{y}_i - y_{i|i-1}(\theta), \qquad 1 \le i \le N \quad ,$$

in (2.2a) and (2.2b) we obtain, by application of the change of variables formula,

$$f(y_1^N;u,\theta) = \prod_{i=1}^{N} f(y_i - \hat{y}_{i|i-1}(\theta)|y_i^{i-1};u,\theta)$$

$$\overline{\nabla} \prod_{i=1}^{N} f_{\theta,u}(y_i - \hat{y}_{i|i-1}(\theta)) \tag{2.3a}$$

and, assuming that the joint density exists,

$$f(y_1^N,u_1^N;\theta) = \prod_{i=1}^{N} f(y_i - \hat{y}_{i|i-1}(\theta)y_1^{i-1},u_1^i;\theta)f(u_i|y_1^{i-1},y_1^{i-1})$$

$$\overline{\nabla} \prod_{i=1}^{N} f_\theta(y_i - y_{i|i-1}(\theta))f(u_i|y_1^{i-1},u_1^{i-1}) \tag{2.3b}$$

respectively, since the determinant of the Jacobian of the change of variables is the identity matrix in each case. (Note that this maneuver works for any change of variables of the form $y_i \to y_i - k_{i-1}(y_1^{i-1})$, $1 \le i \le N$, with $k_{i-1}(\cdot)$ measurable with respect to the sigma field $F\{y_1^{i-1}\}$.)

Let us assume that in (2.3a) y_1^N is a Gaussian random variable and that in (2.3b) (y_1^N,u_1^N) is similarly a Gaussian random variable. Now, $y_i - \hat{y}_{i|i-1}(\theta)$ is necessarily orthogonal to functions which are measurable with respect to $\{y_1^{i-1}\}$, but the addition of the Gaussian assumption makes it *independent* of them. Of course both statements require the data to be distributed according to θ, but this is precisely the assumption in force when we construct the likelihood function. As a result (2.3a) and (2.3b) yield

$$\log f(y_1^N;u,\theta) = \sum_{i=1}^{N} \log f_{\theta,u}(y_i - \hat{y}_{i|i-1}(\theta)) \tag{2.4a}$$

and

$$\log f(y_1^N,u_1^N;\theta) = \sum_{i=1}^{N} \log f_{\theta}(y_i - \hat{y}_{i|i-1}(\theta)) + J(y_1^N,u_1^N) \tag{2.4b}$$

respectively, where $J(y_1^N,u_1^N) \triangleq \log \prod_{i=1}^{N} f(u_i|u_1^{i-1},u_1^{i-1})$.

To summarize, both the assumption (i) y has a Gaussian distribution with (u,θ) deterministic, and (ii) (y,u) has a Gaussian distribution with θ deterministic and $f(u_k|y_k^{k-1},u_k^{k-1};\theta)$ independent of θ for all $k \in \mathbb{Z}_1$, yield a likelihood function of the observations of the form

$$\sum_{i=1}^{N} \log f_{\theta}(\nu_i(\theta)) + \{\text{function of } (y_1^N,u_1^N) \text{ not depending upon } \theta\} \tag{2.5}$$

when $\nu_i(\theta) \triangleq y_i - \hat{y}_{i|i-1}(\theta)$, $1 \le i \le N$, and where $\hat{y}_{i|i-1}(\theta)$ denotes the conditional expectation of y_i given the observations, this being computed using the appropriate conditional distribution in each case.

We now postulate some conditions on the processes appearing in the system equations (1.6) so that the likelihood function on the observations will take the form (2.5) and such that (2.5) will be computable in a convenient recursive manner. These conditions are typical of those imposed on ARMAX and SSX systems to yield a satisfactory theory of filtering and stochastic optimal control.

The following condition will be taken in conjunction with INP 1A and INP 2. Let $\bar{\delta}$ denote the maximum of the row degrees of $A(z)$ in (1.6(i)). Unless $A_\psi(z)$ is row-reduced at each $\psi \in \Psi$, it is in general the case that the sum of the row degrees of $A_\psi(z)$ exceeds the Smith-McMillan degree of $[Z(z)W(z)]$. In the scalar case, of course, $\delta = \bar{\delta}$.

<u>INP 3A</u>. The random variables $\bar{x} \triangleq (y^o_{-\bar{\delta}+1},w^{-1}_{-\bar{\delta}+1})$ and x_0, in ARMAX (1.6a) and SSX (1.6b), respectively, are jointly distributed with and orthogonal to w_0^N for each $N \in \mathbb{Z}_+$, and the joint distribution is Gaussian with zero mean.

In this first (A) case the process u enters the recursions (1.6) in a deterministic manner and the Gaussian density on the respective initial conditions propagates to a Gaussian distribution on $(y^N_{-\bar{\delta}+1},w^N_{-\bar{\delta}+1})$ parameterized by $\theta \equiv (\psi,\Sigma)$. This gives (y_1^N) a (nonsingular) Gaussian distribution for each $N \in \mathbb{Z}_1$.

When INP 1B and INP 2 are in force, we enunciate condition

INP 3B

$(\bar{\bar{x}} \triangleq (y_{-\bar{\delta}+1}^{-1}, u_{-\bar{\delta}+1}^{-1}, w_{-\bar{\delta}+1}^{-1}), (u_0,w_0),(u_1,w_1),\ldots)$ is a zero mean Gaussian process with $w_{n+\tau}$ orthogonal to $Sp\{\bar{\bar{x}},(u_0,w_0),\ldots,(u_n,w_n),u_{n+1}\}$ for all $n \in \mathbb{Z}_+$, $\tau \in \mathbb{Z}_1$. Further, $f(u_k|w_{k-\tau}^{k-1},u_{k-\tau}^{k-1};\theta)$ is not a function of $\theta \in \Theta$ for any $k \in \mathbb{Z}$, $\tau \in \mathbb{Z}_1$.

In this second (B) case, the first hypothesis in INP 3B makes the disturbance w_k, entering the system at the instant k, independent of all preceeding and simultaneous events. In particular, this rules out instantaneous or anticipative feedback from the output process y to the input process u.

The interpretation of the second hypothesis in INP 3B is that any feedback regulator making u dependent upon past inputs and outputs is not parameterized by θ.

Let us continue under the assumption that a joint density exist for (y,u). This will be eventually be guaranteed by the addition of hypothesis (4) in INP B in Section 3.

If the observed input process u is absent, INP 3A and 3B become the single hypothesis that the initial conditions for (1.6) and the (full rank) orthogonal process $(w_1,w_2,\ldots)$ are jointly Gaussian, mutually orthogonal and have zero mean.

Since INP 1-3 give a likelihood function of the form (2.5) we obtain the following expression for the $\theta = (\psi,\Sigma)$ dependent part of the logarithm of the likelihood function (when it is scaled by $-\frac{2}{N}$):

$$-\frac{2}{N}\sum_{i=1}^{N} \log\{\frac{1}{(2\pi)^{p/2}} \frac{1}{(\det \Sigma_{i|i-1}(\theta))^{1/2}} \exp -\frac{1}{2}\|\nu_i(\theta)\|^2_{[\Sigma_{i|i-1}(\theta)]^{-1}}\} \tag{2.6a}$$

$$= p\log 2\pi + \frac{1}{N}\sum_{i=1}^{N}\{\log\det\Sigma_{i|i-1}(\theta) + \mathrm{Tr}[\nu_i(\theta)\nu_i^T(\theta)\Sigma_{i|i-1}^{-1}(\theta)]\} \tag{2.6b}$$

For convenience we often refer to (2.6)--less the quantity $p \log 2\pi$ --as the *log-likelihood function* and denote it by $L_N(y_1^N;u,\theta)$ in case A and $L_N(y_1^N,u_1^N;\theta)$ in case B i.e. $L_N(y_1^N;u,\theta)$ and $L_N(y_1^N,u_1^N;\theta)$, respectively, are given by

$$\frac{1}{N}\sum_{i=1}^{N}\{\log\det\Sigma_{i|i-1}(\theta) + \mathrm{Tr}[\nu_i(\theta)\nu_i^T(\theta)\Sigma_{i|i-1}(\theta)]\} \tag{2.7}$$

The matrix inverse $\Sigma_{i|i-1}^{-1}(\theta)$ in (2.10) and (2.11) exists for all $i \in \mathbb{Z}_1$ and for all $\theta \in \Theta$ since (1.6b) yields

$$\Sigma_{i|i-1}(\theta) = H_\theta V_{i|i-1}(\theta)H^T + \Sigma_\theta\ , \quad i \in \mathbb{Z}_1$$

where $V_{i|i-1}(\theta)$ denotes the state estimation error covariance and we have $\Sigma_\theta > 0$ for all $\theta \in \Theta$ by the definition of Θ.

This function will be our main object to study in Section 3. The crucial point is that the ARMAX and SSX models together with the INP hypotheses permit the recursive construction of the prediction errors $\{\nu_k(\theta);\ 1 \le k \le N\}$ via the techniques of recursive filtering. If the data is taken to be related by the ARMAX model then, at any $\theta \in \Theta$, the conditional expectations $\{\hat{y}_{k|k-1}(\theta); k \ge 1\}$ are generated by the recursive linear least squares estimation formulae due to Rissanen and Barbosa [Rissanen and Barbosa, 1969; Rissanen, 1973], on the other hand, if we take the data to be generated by the equivalent state space model (1.6b) then the recursions of the Kalman filter perform this task.

The following example is instructive.

EXAMPLE. Consider the construction of the likelihood function for a stationary system which is given by the system

$$x_{k+1} = \theta x_k + w_k \quad , \tag{2.8a}$$

$$y_k = x_k \qquad \forall k \in \mathbb{Z}_+ \tag{2.8b}$$

where $x_k, y_k, w_k \in \mathbb{R}^1$ for $k \in \mathbb{Z}_+$ and where w is an i.i.d. $N(0,\sigma^2)$ process. Evidently the system is in both ARMA and state space form.

Let us assume (2.8) is in steady state, in other words we picture the system as having been in operation since the remote past. In steady state $\Pi_\infty \triangleq Ex_k^2$ satisfies the Lyapunov equation $\Pi_\infty = \theta\, \Pi_\infty\, \theta + \sigma^2$ and for this to have a finite solution we must naturally assume that the system is asymptotically stable i.e. $|\theta| < 1$.

The covariance sequence for the observed process y is

$$R_k = R_{-k} = Ey_{k+\tau}y_\tau = \frac{\theta^k\sigma^2}{1-\delta^2}, \qquad \forall k,\tau \in \mathbb{Z} \quad .$$

We shall let Σ_N denote the covariance of $(y_n,\ldots,y_1)$; hence

$$(\Sigma_N)_{i,j} = \frac{\theta^{|i-j|}\sigma^2}{1-\theta^2} \quad , \qquad 1 \le i,j \le N \quad ,$$

and the Gaussian assumption on w makes the entire y process Gaussian with density

$$f(y_1^N;\theta) = \frac{1}{(2\pi)^{N/2}} \frac{1}{|\Sigma_N|^{1/2}} \exp - \frac{1}{2} [y_N, y_{N-1}, \ldots, y_2, y_1] \times$$

$$\begin{bmatrix} \frac{\sigma^2}{1-\theta^2} & \frac{\theta\sigma^2}{1-\theta^2} & & \frac{\theta^{N-1}\sigma^2}{1-\theta^2} & \\ \frac{\theta\sigma^2}{1-\theta^2} & \frac{\sigma^2}{1-\theta^2} & & & \\ & & \ddots & & \\ & & & \frac{\sigma^2}{1-\theta^2} & \frac{\theta\sigma^2}{1-\theta^2} \\ \frac{\theta^{N-1}\sigma^2}{1-\theta^2} & \cdots & & \frac{\theta\sigma^2}{1-\theta^2} & \frac{\sigma^2}{1-\theta^2} \end{bmatrix}^{-1} \begin{bmatrix} y_N \\ y_{N-1} \\ \vdots \\ y_2 \\ y_1 \end{bmatrix}$$

In this simple example we can carry out matrix manipulations to obtain

$$f(y_1^N;\theta) = (2\pi)^{-N/2} \left(\frac{\sigma^2}{1-\theta^2}\right)^{-1/2}, \quad \exp - \frac{1}{2} [y_N, y_{N-1}, \ldots, y_2, y_1] \times$$

$$\begin{bmatrix} 1 & & & & 0 \\ -\theta & 1 & & & \\ & -\theta & 1 & & \\ & & \ddots & \ddots & \\ & & & \ddots & 1 \\ 0 & & & -\theta & 1 \end{bmatrix}$$

$$\times \begin{bmatrix} \sigma^2 & & & & \\ & \sigma^2 & & & \\ & & \ddots & & \\ & & & \sigma^2 & \\ 0 & & & & \frac{\sigma^2}{1-\theta^2} \end{bmatrix}^{-1} \begin{bmatrix} 1 & -\theta & & & 0 \\ & 1 & -\theta & & \\ & & 1 & \ddots & \\ & & & 1 & -\theta \\ 0 & & & & 1 \end{bmatrix} \begin{bmatrix} y_N \\ y_{N-1} \\ \vdots \\ y_2 \\ y_1 \end{bmatrix}$$

Hence

$$-\log f(y_1^N;\theta) = \frac{N}{2}\log 2\pi + \frac{1}{2}\left\{\log\left(\frac{\sigma^{2N}}{1-\theta^2}\right)\right\} + \frac{1}{2}\left\{y_1^2\left(\frac{\sigma^2}{1-\theta^2}\right)^{-1}\right.$$

$$\left. + \sum_{i=2}^{N}(y_i - \theta y_{i-1})^2\sigma^{-2}\right\} \tag{2.9}$$

On the other hand we can obtain (2.9) in the form (2.7) as follows:

$$-\log f(y_1^N;\theta) = \frac{N}{2}\log 2\pi + \frac{1}{2}\sum_{i=1}^{N}\log \Sigma_{i|i-1} + \frac{1}{2}\sum_{i=1}^{N}(y_i-\hat{y}_{i|i-1})^2(\Sigma_{i|i-1})^{-1} \tag{2.10}$$

where $\Sigma_{i|i-1} = E(y_i - \hat{y}_{i|i-1})^2$.

Since $Ew_k = 0$, for all $k \in \mathbb{Z}$, and since no observations are taken before $i = 1$, we have

$$\hat{y}_{1|0} = Ey_1 = 0$$

and

$$\Sigma_{1|0} = E(y_1 - \hat{y}_{1|0})^2 = Ey_1^2 = \Pi_\infty \quad ,$$

when Π_∞ satisfies the Lyapunov equation $\Pi_\infty = \theta\Pi_\infty\theta + \sigma^2$ with solution $\Pi_\infty = \frac{\sigma^2}{1-\theta^2}$. Subsequently $\Sigma_{i|i-1}$ satisfies the Riccati equation

$$\Sigma_{i+1|i} = \theta^2\Sigma_{i|i-1} - \frac{\theta^2\Sigma_{i|i-1}^2}{\Sigma_{i|i-1}} + \sigma^2 = \sigma^2 \ , \quad i \in \mathbb{Z}_+ \quad .$$

Hence the solution sequence to the Riccati equation (with its initial condition determined by the Lyapunov equation) is

$$\frac{\sigma^2}{1-\theta^2}, \sigma^2, \sigma^2, \ldots \quad .$$

and (2.10) is given explicitly by

$$-\log f(y_1^N;\theta) = \frac{N}{2}\log 2\pi + \frac{1}{2}\left\{\log\left(\frac{\sigma^2}{1-\theta^2}\right) + (N-1)\log\sigma^2\right\}$$

$$+ \frac{1}{2}\left\{y_1^2\left(\frac{\sigma^2}{1-\theta^2}\right) + \sum_{i=2}^{N}(y_i - \hat{y}_{i|i-1})^2\sigma^2)\right\}$$

which is the same as (2.8) since

$$y_i - \hat{y}_{i|i-1} = y_i - \theta y_{i-1}$$

for all $i \geq 2$.

This simple example shows how the calculation of $(y_1^N)^T(\Sigma_N)^{-1}(y_1^N)$, in the direct version of the likelihood function, is transformed into a sum of quantities that depend upon the solution of the appropriate Lyapunov and Riccati equations, in the "filter" form of the likelihood function (2.7). This illustrates how, at the cost of the complexity of the Riccati equation (equivalently the Cholesky factorization algorithms), the direct likelihood function as a Gaussian process generated by the system (1.6a), (1.6b) is transformed into the versions depending upon the process innovations.

3. CONSISTING AND ASYMPTOTIC NORMALITY OF MAXIMUM LIKELIHOOD ESTIMATORS: COMPACT PARAMETER SETS. In this section we present our main result; it is that maximum likelihood estimators of the parameters of Gaussian ARMAX and state space systems are strongly consistent and asymptotically normal when the parameters lie in a compact set. The methods of this section are based on techniques which originally appeared in the papers Caines and Rissanen [1974] and Rissanen and Caines [1979].

The analysis will be carried out without making a distinction between the autoregressive moving average models of (1.6a) and the state space models of (1.6b); in fact we shall exploit the equivalence of these models by referring to one or the other according to convenience. On the other hand, there are some significant differences between the analysis in the case where the input u is deterministic, corresponding to our input hypotheses labelled A, and those where it is random, corresponding to our hypotheses labelled B. In this paper we only treat the first case in any detail.

We emphasize here that one of the main advantages in having available a version of the theorem established under input hypotheses B is that it covers the case of time invariant, stationary output to stationary input, feedback, when this feedback relation does not depend upon θ.

In the theorem statement below we gather together, for ease of reference, all the relevant input hypotheses. The reader should note the addition of the important "persistent excitation" condition in both sets of hypotheses: this requires the exogenous input, whether deterministic or random, to behave like a wide sense stationary process which is not linearly deterministic with respect to regressions of order less than twice an integer depending upon the system structure.

In this paper we only present an outline of the proof of the main result for input hypotheses A, omitting the proofs of several of the main technical lemmas and the entire analysis under hypotheses B; these are to be found in the forthcoming text by Caines (see the references at the end of this paper). In several instances, the missing proofs resemble those of earlier results in the papers previously referred to by Rissanen and the author. However, in the case of Lemmas 3.3 and 3.5, they are thought to be of sufficient novelty to merit inclusion in this paper.

Finally, before starting the main theorem, we remark that it is *not* hypothesized that the observed y process $((y,u)$ process, respectively) is stationary. However, because the system is assumed asymptotically stable and the process w, $((u,w)$, respectively) is assumed strictly stationary, the process y $((y,u)$, respectively) is asymptotically stationary in the sense that the covariance of quantities separated by τ instants converges to a quantity depending only upon τ.

THEOREM 3.1. Part 1. *With the* $p,m,\delta \in \mathbb{Z}_+$ *of the output, exogenous input and state processes, respectively, given, consider the ARMAX systems*

$$A_\psi(z)y(z) = B_\psi(z)u(z) + C_\psi(z)w(z) \tag{1.6a}$$

in $S(p,m,\delta)$ *evolving on* $\mathbb{Z}_1$, *or, equivalently, the state space systems*

$$x_{k+1} = F_\psi x_k + G^u_\psi u_k + G^w_\psi w_k \tag{1.6b(i)}$$

$$y_k = H_\psi w_k + D^u_\psi u_k + w_k \tag{1.6b(ii)}$$

in $S(p,m,\delta)$ *evolving on* $\mathbb{Z}_1$. *As indicated, let members of* $S(p,m,\delta)$ *be parameterized by the elements* Ψ *of the (analytic) parameterizing manifold* Ψ; *further, let* $P \,\underline{\Delta}\, \{\Sigma : \Sigma = \Sigma^T > 0,\ \Sigma \in \mathbb{R}^{p^2}\}$ *parameterize the covariance matrices of orthogonal input process* w *and let* $\Theta \,\underline{\Delta}\, \Psi \times P$.

Let the initial conditions for the system and the system inputs u *and* w *be defined on the probability space* $(\Omega,\mathcal{B},P)$ *and let one of the following two alterantive sets of assumptions hold:*

INP A.

(1) u *is deterministic (i.e.* $\{\phi,\Omega\}$ *measurable) and bounded;*

(2) w *is a zero mean wide sense stationary process with* $Ew_k w_j^T = \Sigma\delta_{kj}$, $k,j \in \mathbb{Z}$, *with* $\Sigma \in P$;

(3) $\bar{x} = (y^0_{-\bar{\delta}+1}, w^{-1}_{-\bar{\delta}+1})$ *and* x_0, *in (1.6a) and (1.6b), respectively, are jointing distributed with and orthogonal to* w_0^N *for each* $N \in \mathbb{Z}_+$ *and the joint distribution is Gaussian with zero mean;*

(4) *for the process* u *the following limits exist:*

$$\lim_{N\to\infty} \frac{1}{N} \sum_{i=1}^{N} u_i u_{i-k}^T = M_k, \quad k \in \mathbb{Z}$$

and

$$\begin{bmatrix} M_0 & M_1 & \cdots & M_{2\bar{\delta}} \\ M_1^T & M_0 & & \\ \vdots & & & \\ M_{2\bar{\delta}}^T & \cdots & & M_0 \end{bmatrix} > 0 \quad , \tag{3.1}$$

where this property is referred to as *(deterministic) persistent excitation of order* $2\bar{\delta}$;

(5) *the limit*

$$\lim_{N\to\infty} \frac{1}{N} \sum_{i=1}^{N} u_i = \bar{u}$$

exists.

When the input hypotheses INP A(1-4) are in force the observations on a system in $S(p,m,\delta)$ *consist of the entire sample path on* $\mathbb{Z}$ *of the deterministic observed process* u *and the sample path of* y *on* $\mathbb{Z}_1$.

INP B.

(1) u *is a full rank strictly stationary ergodic process;*

(2) w *is a zero mean wide sense stationary process with* $Ew_k w_j^T = \Sigma\delta_{kj}$, $k,j \in \mathbb{Z}$, *with* $\Sigma \in P$;

(3) $(\bar{\bar{x}} \underline{\Delta} (y^0_{-\bar{\delta}+1}, u^{-1}_{-\bar{\delta}+1}, w^{-1}_{-\bar{\delta}+1}),\ (u_0,w_0),(u_1,w_1),\ldots)$ *is a zero mean Gaussian process with* $w_{n+\tau}$ *orthogonal to* $Sp\{\bar{\bar{x}},(u_0,w_0),\ldots,(u_n,w_n),u_{n+1}\}$ *for any* $n \in \mathbb{Z}_+$, $\tau \in \mathbb{Z}_1$. *Further* $f(u_k|w^{k-1}_{k-\tau},u^{k-1}_{k-\tau};\theta)$ *is not a function of* $\theta \in \Theta$ *for any* $k \in \mathbb{Z}$, $\tau \in \mathbb{Z}_1$;

(4) *let* $v = \{v_k \underline{\Delta} u_k - (u_k|\underline{H}^w_k);k \in \mathbb{Z}\}$, *and call this the* *exogenous part* *of the input process. Then the limits*

$$\lim_{N\to\infty} \frac{1}{N} \sum_{i=1}^{N} v_i v_{i-k}^T = M_k \ , \quad k \in \mathbb{Z}$$

exist, and

$$\begin{bmatrix} M_0 & M_1 & \cdots & M_{2\bar{\delta}} \\ M_1^T & M_0 & & \\ \vdots & & & \\ M_{2\bar{\delta}}^T & \cdots & \cdots & M_0 \end{bmatrix} > 0 \quad , \tag{3.1b}$$

and u *is referred to as a* <u>*(random) persistently exciting process of order*</u> $2\bar{\delta}$.

When the input hypotheses INP B are in force the observations on a system in $S(p,m,\delta)$ *consist of the sample paths of* (y,u) *on* $\mathbb{Z}_1$.

For a system in $S(p,m,\delta)$ *parameterized by* $\theta \in \Theta$, *and generating the observed process subject to the assumptions INP A or INP B, let the maximum likelihood estimate of* θ *be given by a (uniquely specified) maximizing argument for* $L_N(y_1^N;u,\theta)$, *or* $L_N(y_1^N,u_1^N;\theta)$, *respectively, over* $\theta \in \Theta$, *where* C *is a compact subset of* Θ *containing* $\overset{\circ}{\theta}$.

Then $\hat{\theta}_N$ *is strongly consistent, that is to say,*

$$\hat{\theta}_N \to \overset{\circ}{\theta} \quad \text{a.s.} \quad \text{as} \quad N \to \infty$$

in the topology of Θ.

<u>Part 2</u>. *Let the hypotheses INP A or INP B hold. In addition, assume the parameter* $\overset{\circ}{\theta}$ *only takes values in the interior* $\overset{\circ}{C}$ *of* C. *Then*

$$\sqrt{N}\,(\hat{\theta}_N - \overset{\circ}{\theta}) \xrightarrow{\text{dist}} N(0,H(\overset{\circ}{\theta})^{-1}M(\overset{\circ}{\theta})H(\overset{\circ}{\theta})^{-1}) \quad ,$$

where

(i) $\hat{\theta}_N - \overset{\circ}{\theta}$ *denotes the vector difference of* $\hat{\theta}_N$ *and* $\overset{\circ}{\theta}$ *in a local co-ordinate chart of* Θ *around* $\overset{\circ}{\theta}$, *this vector difference necessarily being defined for all suitably large* N, *and*

(ii)

$$[H(\overset{\circ}{\theta})]_{ij} = \frac{1}{2\pi}\frac{\partial^2}{\partial\theta_i\partial\theta_j}\{\log(\det \Sigma(\theta)) + \mathrm{Tr}[\Phi_{\overset{\circ}{\theta}}(\theta) + \Omega_{\overset{\circ}{\theta}}(\theta)]\Sigma^{-1}(\theta)\}\Big|_{\theta=\overset{\circ}{\theta}} \qquad 1 \le i,j \le n \quad , \tag{3.2}$$

where the indicated matrix inverse and second partial differentials necessarily exist, where H *is block diagonal with blocks corresponding to* Σ *and* θ *and where*

$$\Phi_{\overset{\circ}{\theta}}(\theta) \triangleq \frac{1}{2\pi}\int_0^{2\pi} W_\theta^{-1}(e^{i\lambda})W_{\overset{\circ}{\theta}}(e^{i\lambda})\Sigma(\theta)W_{\overset{\circ}{\theta}}^T(e^{-i\lambda})W_\theta^{-T}(e^{-i\lambda})d\lambda$$

and

$$\Omega_{\overset{\circ}{\theta}}(\theta) \triangleq \frac{1}{2\pi}\int_0^{2\pi} W_\theta^{-1}(e^{i\lambda})\tilde{Z}_{\overset{\circ}{\theta}}(e^{i\lambda})dF_u(e^{i\lambda})\tilde{Z}_\theta^T(e^{-i\lambda})W_\theta^{-T}(e^{-i\lambda})d \quad ;$$

where

$$\tilde{Z}_{\overset{\circ}{\theta}}(z) \triangleq Z_\theta(z) - Z_{\overset{\circ}{\theta}}(z)$$

and where F_u *is the non-decreasing function in* $[0,2\pi]$, *such that* $M_k = \frac{1}{2\pi}\int_0^{2\pi} e^{-ik\theta}dF_u(\theta)$, *for* $k \in \mathbb{Z}$, *where this function necessarily exists in both cases A and B.*

Formulae for the blocks of $M(\overset{\circ}{\theta})$ *under input hypotheses A are given in (3.37) and (3.38) below. Under both input hypotheses A and B,* $M(\overset{\circ}{\theta})$ *is block diagonal with blocks corresponding to* Σ *and* θ. *It follows that the covariance matrix* $H(\overset{\circ}{\theta})^{-1}M(\overset{\circ}{\theta})H(\overset{\circ}{\theta})$ *of the limiting distribution of* $\sqrt{N}\,(\hat{\theta}_N - \overset{\circ}{\theta})$ *is similarly block diagonal.*

Outline of Proof. Strong Consistency Under Input Hypotheses A. The proof requires a chain of lemmas which are proved under various explicitly stated hypotheses. When the conditions of a lemma are declared to include "the general hypotheses of the theorem" this indicates that all the conditions of ARMAX 1 and SSX 1 which give rise to the parameterization of $S(p,m,\delta)$ by Θ are in force, and that the observed processes are generated by the system parameterized by $\overset{\circ}{\theta} \in \Theta$. The conditions on the input processes are specified separately.

The condition for part 1 of the theorem, that $\overset{\circ}{\theta} \in C$ (a compact set in Θ, and for part 2, that $\overset{\circ}{\theta} \in \overset{\circ}{C}$ (the interior of C), are not required as conditions for any of the lemmas.

We begin by defining, as in the theorem statement, the process

$$\begin{aligned}\nu^\infty(\theta)(z) &\triangleq C^{-1}(z)[A_\theta(z)y(z) - B_\theta(z)u(z)] \\ &= C_\theta^{-1}(z)A_\theta(z)[(A_{\overset{\circ}{\theta}}^{-1}(z)B_{\overset{\circ}{\theta}}(z)-A_\theta^{-1}(z)B_\theta(z)u(z)+A_{\overset{\circ}{\theta}}^{-1}(z)C_{\overset{\circ}{\theta}}(z)w(z)]\end{aligned} \tag{3.3}$$

for each $\theta \in \Theta$. By ARMAX 1 (equivalently, SSX 1), the rational transfer functions in this expression are asymptotically stable for each $\theta \in \Theta$. Consequently, by INP A(1), the first deterministic part of $\nu^\infty(\theta)$ is a sequence with values in $\mathbb{R}^m$ and it may be shown that the second stochastic part

$$\eta_\theta(z) \triangleq [C_\theta^{-1}A_{\overset{\circ}{\theta}}(z)](A_{\overset{\circ}{\theta}}^{-1}(z)C_\theta(z)]w(z) \underset{\nabla}{=} W_\theta^{-1}(z)W_{\overset{\circ}{\theta}}(z)w(z) \tag{3.4}$$

is a wide sense stationary zero mean Gaussian process. Since the covariance sequence $\{E\eta_{\theta,k}\eta^T_{\theta,k-\tau};\ \tau \in \mathbb{Z}\}$ of the process η_θ is evidently summable, η_θ is a strictly stationary ergodic process.

We wish to study the asymptotic properties of a steady state version of $L_N(y_1^N;u,\theta)$. In order to do this we define the function

$$L_N(\theta) = \log(\det \Sigma(\theta)) + \frac{1}{N}\sum_{i=1}^{N} \mathrm{Tr}[\nu_i^\infty(\theta)\nu_i^{\infty^T}(\theta)\Sigma(\theta)]\}, \quad N \in \mathbb{Z}_1 \ . \tag{3.5}$$

over Θ.

We present, as a self contained lemma, the following fact:

LEMMA 3.1. *Subject to the general hypotheses of the theorem and subject to hypotheses INP A(1-4)*

$$L_N(\theta) \to L_{\overset{o}{\theta}}(\theta) \qquad \text{a.s.} \tag{3.6}$$

as $N \to \infty$ *for all* $\theta \in \Theta$, *moreover this convergence is uniform over any compact set* $D \subset \Theta$.

In (3.6) the function $L_{\overset{o}{\theta}}(\theta)$ *is given by*

$$L_{\overset{o}{\theta}}(\theta) = \log(\det \Sigma(\theta)) + \mathrm{Tr}[[\Phi_{\overset{o}{\theta}}(\theta) + \Omega_{\overset{o}{\theta}}(\theta)]\Sigma^{-1}(\theta)] \tag{3.7)(i}$$

where

$$\Phi_{\overset{o}{\theta}}(\theta) \triangleq E\eta_o(\theta)\eta_o^T(\theta) = \frac{1}{2\pi}\int_0^{2\pi} W_\theta^{-1}(e^{i\lambda})W_{\overset{o}{\theta}}(e^{i\lambda})\Sigma(\overset{o}{\theta})W_{\overset{o}{\theta}}^T(e^{-i\lambda})W_\theta^{-T}(e^{i\lambda})d\lambda \tag{3.7)(ii)}$$

$$\Omega_{\overset{o}{\theta}}(\theta) \triangleq \frac{1}{2\pi}\int_0^{2\pi} W_\theta^{-1}(e^{i\lambda})\tilde{Z}_\theta(e^{i\lambda})dF_u(e^{i\lambda})\tilde{Z}_\theta^T(e^{-i\lambda})W_\theta^{-T}(e^{-i\lambda}) \ , \tag{3.7)(iii}$$

$$\tilde{Z}_\theta(z) \triangleq Z_\theta(z) - Z_{\overset{o}{\theta}}(z) \tag{3.7)(iv}$$

and where $F_u(\cdot)$ *is a non-decreasing function on* $[0,2\pi]$ *such that*

$$M_k = \frac{1}{2\pi}\int_0^{2\pi} e^{-ik\lambda}dF_u(\lambda) \ , \qquad \forall k \in \mathbb{Z} \ ,$$

this function necessarily existing.

We remark immediately that F_u exists since, for all N, the $Nm \times Nm$ matrix, with i,j-th block entry M_{i-j}, is positive semi-definite and hence we may apply the theorem of Herglotz to produce the required function. The next fact is that the exact likelihood function

$$L_N(y_1^N;u,\theta) = \frac{1}{N}\sum_{i=1}^{N} \{\log \det \Sigma_{i|i-1}(\theta) + \mathrm{Tr}[\nu_i(\theta)\nu_i^T(\theta)\Sigma_{i|i-1}(\theta)] \tag{2.7}$$

converges a.s. and uniformly to its steady state version $L_N(\theta)$. This will also be given as a self contained lemma:

LEMMA 3.2. *Subject to the general hypotheses of the theorem and subject to general input hypotheses INP A(1)-(4).*

$$L_N(y_1^N;u,\theta) \to L_N(\theta) \qquad \text{a.s.} \tag{3.8}$$

as $N \to \infty$ *for all* $\theta \in \Theta$ *and uniformly over any compact set* $D \subset \Theta$.

We are now in a position to prove the strong consistency part of the theorem under the input hypotheses INP A.

In order to apply Lemma 3.1 and 3.2 we identify the compact set D with the set C containing $\overset{o}{\theta}$.

By the definition of $\hat{\theta}_N$

$$L_N(y_1^N;u,\theta) \geq L_N(y_1^N;u,\hat{\theta}_N) \qquad \text{a.s.} \tag{3.9}$$

for each $N \in \mathbb{Z}_1$ and all $\theta \in C$. Not C is a compact subset of a topological manifold and so it is sequentially compact (see e.g. Kelley [1955]). Consequently the sequence $\{\hat{\theta}_N;N \in \mathbb{Z}_1\}$ has a convergent subsequence $\{\hat{\theta}_{N_M};M \in \mathbb{Z}_1\}$ such that $\hat{\theta}_{N_M} \to \theta^*$ in the topology of Θ as M tends to ∞.

We observe that θ^* is a $\mathcal{B}$ measurable Θ valued random variable.

The next step is to insert θ^* in the left hand side of (3.9) and take limits along the sequence $\{N_M;M \in \mathbb{Z}_1\}$. Then by Lemma 3.1 and 3.2

$$L(\overset{o}{\theta}) \geq L(\theta^*) \qquad \text{a.s.} \quad .$$

The details are as follows:

For any $\varepsilon > 0$, Lemma 3.1 assures us of the existence of $N_1(\omega,\varepsilon) \in \mathbb{Z}_1$ such that, for all $N > N_1(\omega,\varepsilon)$,

$$|L_N(\theta) - L_{\overset{o}{\theta}}(\theta)| < \varepsilon/3 \qquad \text{a.s.,} \qquad \forall\theta \in C \quad ; \tag{3.10}$$

Next, Lemma 3.2 guarantees the existence of $N_2(\omega,\varepsilon) \in \mathbb{Z}_1$ such that, for all $N > N_2(\omega,\varepsilon)$,

$$|L_N(y_1^N;u,\theta) - L_N(\theta)| < \varepsilon/3 \qquad \text{a.s.,} \qquad \forall\theta \in C \quad , \tag{3.11}$$

where we have used the uniformity of the convergence in both (3.10) and (3.11).

Combining (3.9), (3.10) and (3.11) yields

$$L_{\overset{o}{\theta}}(\theta) > L_N(y_1^N;u,\theta) - 2\,\varepsilon/3 \geq L_N(y_1^N;u,\hat{\theta}_N) - 2\varepsilon/3 \quad \text{a.s.} \tag{3.12}$$

for $N > \max(N_1\ N_2)(\omega,\varepsilon)$ for all $\theta \in C$.

Finally, pick $N_3(\omega,\varepsilon)$ so that, for all M such that $N_M > N_3(\omega,\varepsilon)$, $\hat{\theta}_{N_M}$ lies in the neighbourhood $D_{\theta^*}(\varepsilon)$ of θ^* for which

$$|L_{\overset{\circ}{\theta}}(\theta) - L_{\overset{\circ}{\theta}}(\theta^*)| < \varepsilon/3 \qquad \forall\theta \in D_{\theta^*}(\varepsilon) \quad , \tag{3.13}$$

where this neighbourhood exists by the continuity of $L_{\overset{\circ}{\theta}}(\cdot)$ on Θ (see (3.7)).

From (3.10) and (3.13) evaluated at $\theta = \hat{\theta}_{N_M}$ we conclude that

$$|L_{N_M}(y_1^{N_M},u,\hat{\theta}_{N_M}) - L_{\overset{\circ}{\theta}}(\theta^*)| < \varepsilon \qquad \text{a.s.}$$

for all M such that $N_M > \max(N_1,N_2,N_3)(\omega,\varepsilon)$. Combining the relation with (3.12), evaluated at $\theta = \overset{\circ}{\theta}$, yields

$$L_{\overset{\circ}{\theta}}(\overset{\circ}{\theta}) > L_{\overset{\circ}{\theta}}(\theta^*) - 5\,\varepsilon/3 \qquad \text{a.s.} \tag{3.14}$$

for all $N_M > \max(N_1,N_2,N_3)(\omega,\varepsilon)$. But this expression is independent of N_M, and ε is arbitrary. So (3.14) implies that

$$L_{\overset{\circ}{\theta}}(\overset{\circ}{\theta}) \geq L_{\overset{\circ}{\theta}}(\theta^*) \qquad \text{a.s.} \tag{3.15}$$

However, in Lemma 3.3 below it is shown that $L_{\overset{\circ}{\theta}}(\theta) \geq L_{\overset{\circ}{\theta}}(\overset{\circ}{\theta})$ a.s. for all $\theta \in C$, with equality holding if and only if $\theta = \overset{\circ}{\theta}$. Hence (3.15) and Lemma 3.3 together give

$$L_{\overset{\circ}{\theta}}(\theta^*) \geq L_{\overset{\circ}{\theta}}(\theta) \geq L_{\overset{\circ}{\theta}}(\theta^*) \qquad \text{a.s.}$$

with $\overset{\circ}{\theta} = \theta^*$ a.s. and we conclude that all subsequent limits of $\{\hat{\theta}_N; N \in \mathbb{Z}_1\}$ equal $\overset{\circ}{\theta}$ a.s. or, equivalently,

$$\hat{\theta}_N \to \overset{\circ}{\theta} \qquad \text{a.s.}$$

as $N \to \infty$. This is the desired strong consistency property of $\{\hat{\theta}_N; N \in \mathbb{Z}_1\}$.

So the only step remaining is to establish

LEMMA 3.3. *Under the general hypotheses of the theorem and the input hypotheses A1-A4*

$$L_{\overset{\circ}{\theta}}(\theta) \geq L_{\overset{\circ}{\theta}}(\overset{\circ}{\theta}) \qquad \text{a.s.,} \qquad \forall\theta \in \Theta \tag{3.16}$$

where, as in (3.7),

$$L_{\overset{\circ}{\theta}}(\theta) = \log\det\Sigma(\theta) + \mathrm{Tr}[E\eta_0(\theta)\eta_0^T(\theta) + \Omega(\theta)]\Sigma^{-1}(\theta) \tag{3.7}$$

Further, equality holds in (3.16) if and only if $\theta = \overset{\circ}{\theta}$.

Proof. We show that $L_{\overset{\circ}{\theta}}(\theta)$ has a unique global minimum at $\theta = \overset{\circ}{\theta}$. Consider the function $\log \det X + \mathrm{Tr}[QX^{-1}]$ where X and Q are $(p \times p)$ positive definite matrices. One may verify that the inequality

$$\log(\det X) + \mathrm{Tr}\, QX^{-1} \geq \log \det Q + p \tag{3.17}$$

holds with the lower bound on the right hand side of (3.17) attained only at $X = Q$.

Identifying terms between (3.7) and (3.17) we obtain the lower bound

$$\log(\det[\Phi_{\overset{\circ}{\theta}}(\theta) + \Omega(\theta)]) + p \tag{3.18}$$

for (3.7). Now the function $\log(\cdot)$ is strictly monotone increasing on $\mathbb{R}^1$ and $\det A > \det B$ for $A \neq B$, $A = A^T \geq B = B^T > 0$. (This follows since $\det(X + I) > \det I$ when $X = X^T \geq 0$ and $X \neq 0$.) Consequently we may obtain a *unique* global minimum for (3.18) at $\psi = \overset{\circ}{\psi}$ if we can show $\Phi_{\overset{\circ}{\theta}}(\theta) + \Omega_{\overset{\circ}{\theta}}(\theta)$ has a unique positive definite global minimum (as a positive semi-definite matrix) at $\psi = \overset{\circ}{\psi}$, i.e., $\Phi_{\overset{\circ}{\theta}}(\theta) + \Omega_{\overset{\circ}{\theta}}(\theta) \geq \Phi_{\overset{\circ}{\theta}}(\overset{\circ}{\theta}) + \Omega_{\overset{\circ}{\theta}}(\overset{\circ}{\theta}) > 0$ with equality only if $\psi = \overset{\circ}{\psi}$. We do this by showing $\Phi_{\overset{\circ}{\theta}}(\theta) \geq \Phi_{\overset{\circ}{\theta}}(\overset{\circ}{\theta}) = \Sigma(\overset{\circ}{\theta}) > 0$ and $\Omega_{\overset{\circ}{\theta}}(\theta) \geq \Omega_{\overset{\circ}{\theta}}(\overset{\circ}{\theta}) = 0$ with equality in both only if $\psi = \overset{\circ}{\psi}$.

$\eta(\theta)$ is the "residual sequence" obtained by applying the θ-parameterized steady state filter to the full rank stochastic part $u = W_{\overset{\circ}{\theta}}w$ of the observed process y. Hence it is intuitively clear that the variance

$$E\eta_0(\theta)\eta_0^T(\theta) = \frac{1}{2\pi}\int_0^{2\pi} W_\theta^{-1}(e^{i\lambda})W_{\overset{\circ}{\theta}}(e^{i\lambda})\Sigma(\overset{\circ}{\theta})W_{\overset{\circ}{\theta}}^T(e^{-i\lambda})W_\theta^{-T}(e^{-i\lambda})d\lambda$$

minimized at $W_\theta(z) = W_{\overset{\circ}{\theta}}(z)$.

By the stability assumptions on systems parameterized by Θ, the sums $W_\theta^{-1}(z) = I + P(\theta,z) = I + \sum_{j=1}^{\infty} P_j(\theta)z^j$ and $W_{\overset{\circ}{\theta}}(z) = I + Q(\overset{\circ}{\theta},z) = I + \sum_{j=1}^{\infty} Q_j(\overset{\circ}{\theta})z^j$ converge absolutely and uniformly to an analytic, and hence bounded, function inside $\{z; |z| \leq 1 + \varepsilon,\ z \in \mathbb{C}\}$ for some $\varepsilon > 0$.

Then, by the orthogonality of the exponential functions on the unit circle (or by using the fundamental result on the prediction of a wide sense stationary process from its own past) we have

$$E\eta_0(\theta)\eta_0{}^T(\theta) = \frac{1}{2\pi}\int_0^{2\pi} (I + P(\theta,e^{i\lambda}))(I + Q(\overset{\circ}{\theta},e^{i\lambda}))\Sigma(\overset{\circ}{\theta})(I + Q(\theta,e^{-i\lambda}))^T$$

$$(I + P(\overset{\circ}{\theta},e^{i\lambda}))^T d\lambda \geq \Sigma(\overset{\circ}{\theta}) \tag{3.19}$$

Now, since $\Sigma(\theta) > 0$ for all $\theta \in \Theta$, equality holds in (3.19) if and only if $W_\theta^{-1}(z)W_{\overset{o}{\theta}}(z) = I$. We conclude that $\Phi_{\overset{o}{\theta}}(\theta)$ is minimized as a positive semi-definite matrix if and only if

$$W_\theta(z) = W_{\overset{o}{\theta}}(z) \tag{3.20}$$

in which case $\Phi_{\overset{o}{\theta}}(\theta) = \Sigma(\overset{o}{\theta})$.

It is clear from the definition of $\Omega(\cdot)$ in (3.7(iii),(iv)) that $\Omega(\theta) \geq \Omega(\overset{o}{\theta}) = 0$ for all $\theta \in \Theta$. We need to show that $\Omega(\theta) = 0$ only if $Z(\theta) = Z(\overset{o}{\theta})$. Let us write $W_\theta^{-1}(z)\tilde{Z}_\theta(z)$ in the left co-prime factorization (m.f.d.) form $F_\theta^{-1}(z)G_\theta(z)$, where $F_\theta(z)$ and $G_\theta(z)$ are polynomial matrices of order less than or equal to $2\bar{\delta}$ and $F_\theta(z)$ has a leading non-singular matrix $F_{0,\theta}$ since $W_\theta^{-1}(z)\tilde{Z}_\theta(z)$ has no pole at $z = 0$.

Now in case $\Omega(\theta) = 0$, we have

$$0 = \frac{1}{2\pi}\int_0^{2\pi} F_\theta^{-1}(e^{i\lambda})G_\theta(e^{i\lambda})dF_u(e^{i\lambda})G_\theta^T(e^{-i\lambda})F_\theta^{-T}(e^{-i\lambda})$$

and, since $F_\theta^{-1}G_\theta dF_u G_\theta^* F^{-*} \geq 0$ and F^{-1} exists, we have $G_\theta dF_u G_\theta^* \equiv 0$, when * denotes conjugate transpose.

Hence

$$0 = \int_0^{2\pi} G_\theta(e^{i\lambda})dF_u(e^{i\lambda})G_\theta^T(e^{-i\lambda}) = \sum_{i,j=0}^{2\bar{\delta}} G_i(\theta)M_{i-j}G_j^T(\theta) \tag{3.21}$$

when

$$G_\theta(z) \triangleq \sum_{i=0}^{2\bar{\delta}} G_i(\theta)z^i \quad .$$

But then the hypothesis that u is persistently existing of order $2\bar{\delta}$ (INP A(3)) yields $G_\theta(z) = 0$, and so

$$Z_\theta(z) = Z_{\overset{o}{\theta}}(z) \quad . \tag{3.22}$$

But by ARMAX 1, or SSX 1, (3.20) and (3.22) imply $\psi = \overset{o}{\psi}$.

At this point we have shown that

$$I_{\overset{o}{\theta}}(\theta) \geq \log(\det\,(\Phi_{\overset{o}{\theta}}(\theta) + \Omega_{\overset{o}{\theta}}(\theta)) + p$$

$$\geq \log\det\,(\Sigma(\overset{o}{\theta})) + p$$

with equality in the second inequality if and only if $\psi = \overset{o}{\psi}$.

Now at $\theta = \overset{o}{\theta}$, $L_{\overset{o}{\theta}}(\theta)$ certainly attains the lower bound $\log(\det \Sigma(\overset{o}{\theta})) + p$. On the other hand, $L_{\overset{o}{\theta}}(\theta) = \log(\det \Sigma(\overset{o}{\theta})) + p$ has just been shown to imply $\psi = \overset{o}{\psi}$ and hence, by (3.7),

$$L_{\overset{o}{\theta}}(\theta) = \log(\det \Sigma(\theta)) + \mathrm{Tr}[\Sigma(\overset{o}{\theta})\Sigma^{-1}(\theta)] = \log(\det (\Sigma(\overset{o}{\theta})) + p \quad .$$

But again using the fact that $\log(\det X) + \mathrm{Tr}\, QX^{-1} \geq \log \det Q + p$ (for $X = X^T > 0$ and $Q = Q^T > 0$), with equality only if $X = Q$, we obtain $\Sigma(\theta) = \Sigma(\overset{o}{\theta})$.

Hence $\theta = \overset{o}{\theta}$ is the unique globally minimizing parameter for $L_{\overset{o}{\theta}}(\theta)$ over Θ.

Asymptotic Normality Under Input Hypotheses A. The asymptotic normality part of the theorem has the additional hypotheses

$$\lim_{N\to\infty} \frac{1}{N} \sum_{k=1}^{N} u_k = \bar{u} \quad \text{and} \quad \overset{o}{\theta} \in \overset{o}{C}$$

By the result just proven $\hat{\theta}_N$ lies a.s. within a co-ordinate neighborhood of θ for all N sufficiently large, i.e., $(\hat{\theta}_N - \overset{o}{\theta}) \in N_\varepsilon(0)$ where $N_\varepsilon(0)$ is an ε-neighborhood of the origin in $\mathbb{R}^\nu$.

Since we may pick ε such that $N_\varepsilon(\overset{o}{\theta}) \subset \overset{o}{C}$ we obtain $\frac{\partial L_N}{\partial \theta_i}(y_1^N;u,\hat{\theta}_N) = 0$, $1 \leq i \leq \eta$ and so the Mean Value Theorem applied to $\frac{\partial L_N}{\partial \theta_i}(y_i^N;u,\theta)$ at θ_N yields

$$\frac{-\partial L_N}{\partial \theta_i}(y_1^N;u,\overset{o}{\theta}) = \frac{\partial^2}{\partial\theta\partial\theta_i} L_N(y_1^N;u,\theta^*_{N,i}) \quad (\hat{\theta}_N - \overset{o}{\theta}) \tag{3.23}$$

for $\theta^*_{N,i}$ an interior point of the line segment $[\hat{\theta}_N, \overset{o}{\theta}]$, for all N sufficiently large, where the differential $\frac{\partial^2}{\partial\theta\partial\theta_i}$ yields a row vector.

We observe that the definition of $L_{\overset{o}{\theta}}(\theta)$ in (3.7(i)) and the analyticity of Θ implies that all derivatives required in this section exist and are continuous.

Now in analogy with Lemma 3.1 we have to analyse the behaviour of the relation (3.23). We begin with

LEMMA 3.4. *Subject to the general hypotheses of the theorem and hypotheses INP A(1)-(4)*

$$\frac{\partial^2}{\partial\theta_i\partial\theta_j} L_N(\theta) \to H_{ij}(\theta) \triangleq \frac{\partial}{\partial\theta_i\partial\theta_j} \{\log(\det \Sigma(\theta)) + \mathrm{Tr}[(\Phi_{\overset{o}{\theta}}(\theta) + \Omega_{\overset{o}{\theta}}(\theta))\Sigma^{-1}(\theta)]\},$$

$$1 \leq i,j \leq \eta \quad , \quad \text{a.s.} \tag{3.24}$$

as $N \to \infty$ *for all* $\theta \in \Theta$, *uniformly over any compact subset* $D \subset \Theta$, *where*

$$L_N(\theta) = \log(\det \Sigma(\theta)) + \frac{1}{N}\sum_{k=1}^{N} \nu_k^{\infty T}(\theta)\Sigma^{-1}(\theta)\nu_k^{\infty}(\theta) , \quad N \in \mathbb{Z}_1 \quad ,$$

and $\Phi_{\overset{o}{\theta}}(\theta)$ *and* $\Omega_{\overset{o}{\theta}}(\theta)$ *are defined in the theorem statement.*

Moreover this convergence is uniform over any compact subset D *of* Θ.

Next we have

LEMMA 3.5. *Subject to the general hypotheses of the theorem and the hypotheses INPA (1-4) the* $(\eta \times \eta)$ *matrix* $H(\overset{o}{\theta})$ *is non-singular.*

Proof. The initial part of this proof parallels that of the analogous Lemma 1 in Goodrich and Caines [1979a] which established the non-singularity of the Hessian $H(\overset{o}{\theta})$ for the case of random vector variables, not processes.

To prove $H(\overset{o}{\theta})$ is non-singular we first obtain more explicit formulae for this matrix.

Differentiation with respect to the elements of Σ. From the relation $\frac{\partial}{\partial\theta_i}\log(\det \Sigma(\theta)) = \mathrm{Tr}[\Sigma^{-1}(\theta)\frac{\partial \Sigma}{\partial\theta_i}(\theta)]$, we obtain

$$\frac{\partial^2}{\partial\theta_i\partial\theta_j}\log(\det \Sigma(\theta)) = \mathrm{Tr}[-\Sigma^{-1}(\theta)\frac{\partial\Sigma}{\partial\theta_i}(\theta)\Sigma^{-1}(\theta)\frac{\partial\Sigma}{\partial\theta_j}(\theta) + \Sigma^{-1}(\theta)\frac{\partial^2\Sigma}{\partial\theta_i\partial\theta_j}(\theta)] \tag{3.25}$$

for the first summand in (3.24), where, as in the remainder of this proof, $1 \le i,j \le \eta$.

Next, differentiating $\Sigma^{-1}(\theta)$ in the second term in (3.24), we obtain the contribution

$$\mathrm{Tr}\{[\Phi_{\overset{o}{\theta}}(\theta) + \Omega_{\overset{o}{\theta}}(\theta)][2\Sigma^{-1}(\theta)\frac{\partial\Sigma}{\partial\theta_i}(\theta)\Sigma^{-1}(\theta)\frac{\partial\Sigma}{\partial\theta_j}(\theta)\,\Sigma^{-1}(\theta) - \Sigma^{-1}(\theta)\frac{\partial^2\Sigma}{\partial\theta_i\partial\theta_j}(\theta)\Sigma^{-1}(\theta)]\} \tag{3.26}$$

where the factor 2 is obtained by exploiting the symmetry of all matrices appearing in the "direct" form of the differential. Notice that there is no contribution from the differentiation of $\Phi_{\overset{o}{\theta}}(\theta) + \Omega_{\overset{o}{\theta}}(\theta)$ with respect to Σ.

From this we can see that the contribution to (3.24) associated with the matrix of terms $\frac{\partial^2\Sigma(\theta)}{\partial\theta_i\partial\theta_j}$ is

$$\mathrm{Tr}\{[-\Phi_{\overset{\circ}{\theta}}(\theta) - \Omega_{\overset{\circ}{\theta}}(\theta) + \Sigma(\theta)][\Sigma^{-1}(\theta)\,\frac{\partial^2\Sigma(\theta)}{\partial\theta_i\partial\theta_j}\,\Sigma^{-1}(\theta)]\}\Big|_{\theta=\overset{\circ}{\theta}} \tag{3.27}$$

However $\Omega_{\overset{\circ}{\theta}}(\overset{\circ}{\theta}) = 0$, because $\tilde{Z}_{\overset{\circ}{\theta}}(z) = 0$, and, from the definition of $\Phi_{\overset{\circ}{\theta}}(\cdot)$, $\Phi_{\overset{\circ}{\theta}}(\overset{\circ}{\theta}) = \Sigma(\overset{\circ}{\theta})$. (As in Lemma 3.3). Consequently (3.27) vanishes.

Now, the sum of terms associated with the products of first differentials of $\Sigma(\theta)$ is

$$\mathrm{Tr}\{[2(\Phi_{\overset{\circ}{\theta}}(\theta) + \Omega_{\overset{\circ}{\theta}}(\theta)) - \Sigma(\theta)][\Sigma^{-1}(\theta)\,\frac{\partial\Sigma(\theta)}{\partial\theta_i} - \Sigma^{-1}(\theta)\,\frac{\partial\Sigma(\theta)}{\partial\theta_i}\,\Sigma^{-1}(\theta)]\}\Big|_{\theta=\overset{\circ}{\theta}}$$

$$= \mathrm{Tr}[\,\frac{\partial\Sigma}{\partial\theta_i}\,(\overset{\circ}{\theta})\Sigma^{-1}(\overset{\circ}{\theta})\,\frac{\partial\Sigma}{\partial\theta_j}\,(\overset{\circ}{\theta})\Sigma^{-1}(\overset{\circ}{\theta})]\} \quad , \tag{3.28}$$

where here and below we write $\frac{\partial f}{\partial\theta}(\overset{\circ}{\theta})$ for $\frac{\partial f}{\partial\theta}(\theta)\Big|_{\theta=\overset{\circ}{\theta}}$ with, we trust, no risk of confusion. We conclude that (3.28) is the part of (3.24) associated with second differentials with respect to Σ.

<u>Differentiation with respect to both</u> Σ <u>and</u> $[Z\ W]$. Since $\theta = (\Sigma,\psi)$ separately paramterizes $\Sigma(\theta)$ and $[Z_\theta(z),W_\theta(z)]$ the second mixed partial derivatives all vanish. This is because $\tilde{Z}_{\overset{\circ}{\theta}} = 0$ and for the reason invoked below for the disappearance of the second two terms in (3.30).

In order for $H(\overset{\circ}{\theta})$ to be singular it is necessary and sufficient that there exist $\mu^T = (\mu_1,\mu_2,\ldots,\mu_\eta) \neq 0$ such that $\mu^T H(\theta)\mu = 0$; from (3.28) and the block diagonality $H(\theta)$, which follows from the vanishing of the mixed partials with respect to Σ and ψ, we see that

$$\mathrm{Tr}\left[\sum_{i=1}^{\eta'}\sum_{j=1}^{\eta'} \mu_i\Sigma^{-1/2}(\overset{\circ}{\theta})\,\frac{\partial\Sigma}{\partial\theta_i}\,(\overset{\circ}{\theta})\Sigma^{-1}(\overset{\circ}{\theta})\,\frac{\partial\Sigma}{\partial\theta_j}(\overset{\circ}{\theta})\Sigma^{-1/2}(\overset{\circ}{\theta})\mu_j\right] = 0,$$

$$\eta' = \frac{p(p+1)}{2} \quad ,$$

and this is equivalent to

$$\sum_{i=1}^{\eta'}\mu_i\,\frac{\partial\Sigma}{\partial\theta}\,(\overset{\circ}{\theta}) = 0\ , \qquad \mu \neq 0 \tag{3.29}$$

by the non-singularity of $\Sigma(\overset{\circ}{\theta})$.

<u>Differentiation with respect to elements in</u> $[Z\ W]$. First, we observe there is no contribution from the differentiation of $\log(\det\Sigma(\theta))$. Turning to the second differentials of $\Phi_{\overset{\circ}{\theta}}(\theta)$ and $\Omega_{\overset{\circ}{\theta}}(\theta)$ in (3.24) with respect to the components ψ in $\theta = (\psi,\Sigma)$, we obtain from $\Phi_{\overset{\circ}{\theta}}(\theta)$ the positive semi-definite matrix contribution

$$\mathrm{Tr}\left[\frac{\partial\Phi_{\overset{o}{\theta}}(\theta)}{\partial\theta\,\partial\theta}\Sigma^{-1}(\theta)\right] = \mathrm{Tr}\left[\left\{\frac{2}{2\pi}\int_0^{2\pi}\frac{\partial W_\theta^{-1}}{\partial\theta_i}(e^{i\lambda})W_{\overset{o}{\theta}}(e^{i\lambda})\Sigma(\overset{o}{\theta})W^T(e^{-i\lambda})\frac{\partial W^{-T}}{\partial\theta_j}(e^{-i\lambda})d\lambda\right.\right.$$

$$+\frac{1}{2\pi}\int_0^{2\pi}\frac{\partial^2 W_\theta^{-1}}{\partial\theta_i\partial\theta_j}(e^{i\lambda})W_{\overset{o}{\theta}}(e^{i\lambda})\Sigma(\overset{o}{\theta})W_{\overset{o}{\theta}}^T(e^{-i\lambda})W_\theta^{-T}(e^{-i\lambda})d\lambda \tag{3.30}$$

$$\left.\left.+\frac{1}{2\pi}\int_0^{2\pi}W_\theta^{-1}(e^{-i\lambda})W_{\overset{o}{\theta}}(e^{i\lambda})\Sigma(\overset{o}{\theta})W_{\overset{o}{\theta}}^T(e^{-i\lambda})\frac{\partial^2 W^T}{\partial\theta_i\partial\theta_j}(e^{-i\lambda})d\lambda\right\}\Sigma^{-1}(\overset{o}{\theta})\right] ,$$

$$\frac{p(r+p)}{2} \le i,j \le \eta \quad ,$$

where we have again used the properties of trace to form the first integral.

Now the contribution's last two integrals evaluated at $\theta = \overset{o}{\theta}$ is of the form

$$\mathrm{Tr}\left[\left\{\int_0^{2\pi}\sum_{p=o}^{\infty}A_p e^{ip\lambda}d\lambda + \int_0^{2\pi}\sum_{p=o}^{\infty}A_p^T e^{-ip\lambda}d\lambda\right\}\Sigma^{-1}(\overset{o}{\theta})\right]$$

$$= \mathrm{Tr}\,[(A + (-A^T))\Sigma^{-1}(\overset{o}{\theta})] = 0 \quad ,$$

in an obvious notation, by the properties Tr.

Again if $\mu^T H(\overset{o}{\theta}) = 0$, for some $\mu \neq 0$, then this would imply from (3.30) that

$$\frac{1}{\pi}\int_0^{2\pi}\mathrm{Tr}\, M_\theta(\lambda)M_\theta^{-T}(\lambda)d\lambda\Big|_{\theta=\overset{o}{\theta}} = 0 \quad ,$$

when

$$M_\theta(\lambda) \underline{\Delta} \sum_{i=1}^{\eta}\mu_i^T\frac{\partial W_\theta^{-1}}{\partial\theta_i}(e^{i\lambda})W_{\overset{o}{\theta}}(e^{i\lambda})\Sigma^{-1/2}(\overset{o}{\theta}) \quad .$$

By the continuity of all functions appearing in $M(\cdot)$ on $[0,2\pi]$ this yields

$$\sum_{i=1}^{\eta}\mu_i^T W_{\overset{o}{\theta}}^{-1}(e^{i\lambda})\frac{\partial W_{\overset{o}{\theta}}}{\partial\theta}(e^{i\lambda}) = 0 \quad , \qquad \forall\lambda \in [0,2\pi] \quad . \tag{3.31}$$

The term associated with $\Omega_{\overset{o}{\theta}}(\theta)$ reduces at $\theta = \overset{o}{\theta}$ to

$$\mathrm{Tr}\left[\frac{\partial^2}{\partial\theta_i\partial\theta_j}\,\Omega_{\overset{o}{\theta}}(\overset{o}{\theta})\Sigma^{-1}(\overset{o}{\theta})\right] =$$

$$= \mathrm{Tr}\left[\frac{2}{2\pi}\int_0^{2\pi}\frac{\partial}{\partial\theta_i}\,\{W_\theta^{-1}(e^{i\lambda})\tilde{Z}_\theta(e^{i\lambda})\}dF_u(e^{i\lambda})\,\frac{\partial}{\partial\theta_j}\,\{\tilde{Z}^T(e^{-i\lambda})W_\theta^{-T}(e^{-i\lambda})\}\Sigma^{-1}(\overset{o}{\theta})\right]\Bigg|_{\theta=\overset{o}{\theta}}$$

for the same reasons that (3.30) simplifies. Now using the fact that $\tilde{Z}_{\overset{o}{\theta}}(z) = 0$, and the linearity of trace, the hypothesis $\mu^T H(\overset{o}{\theta})\mu = 0$ yields

$$\mathrm{Tr}\int_0^{2\pi} N_\theta(\lambda)dF_u(e^{i\lambda})N_\theta^{-T}(\lambda)\Bigg|_{\theta=\overset{o}{\theta}} = 0$$

when

$$N_\theta(\lambda) \triangleq \sum_{i=0}^{n}\mu_i W_{\overset{o}{\theta}}^{-1}(e^{i\lambda})\frac{\partial\tilde{Z}_{\overset{o}{\theta}}}{\partial\theta_i}(e^{i\lambda})$$

$$= \sum_{i=0}^{n}\mu_i^T W_{\overset{o}{\theta}}^{-1}(e^{i\lambda})\frac{\partial Z_{\overset{o}{\theta}}}{\partial\theta_i}(e^{i\lambda}) \quad .$$

Using the fact that the row degree of $A(z), B(z), C(z)$ are all less than δ together with the persistent excitation property INP A(4) we obtain

$$\sum_{i=1}^{n}\mu_i^T\,W_\theta^{-1}(e^{i\lambda})\,\frac{\partial\tilde{Z}_\theta}{\partial\theta_i}(e^{i\lambda}) = 0 \quad , \qquad \forall\lambda\in[0,2\pi] \quad . \tag{3.32}$$

(3.29), (3.31) and (3.32) taken together imply

$$\sum_{i=1}^{n}\bar{\mu}_i^T\frac{\partial}{\partial\theta_i}\begin{pmatrix}\Sigma^{[s]}(\overset{o}{\theta}) \\ \\ [Z_{\overset{o}{\theta}}(e^{i\lambda})W_{\overset{o}{\theta}}(e^{i\lambda})]^{[s]}\end{pmatrix} = 0 \quad ,$$

$$\forall\lambda\in[0,2\pi],\ \left(\frac{p(p+1)}{2} + pr + p^2\right)\times 1 \quad , \tag{3.33}$$

where $\bar{\mu}_i = \mu_i$ for the components corresponding to $\Sigma(\theta)$ and $\bar{\mu}_i = \mu^T W_\theta^{-1}(e^{i\lambda})_i$ for the remaining components (recall $H(\theta)$ is block diagonal), $\bar{\mu} \neq 0$, and, for an $m\times n$ matrix A, $A^{[s]}$ denotes the mn vector created by stacking the m-component column vectors of A.

Since (3.33) holds for all $\lambda\in[0,2\pi]$, and since $Z(z)$ and $W(z)$ are rational functions of z,

$$\sum_{i=1}^{\eta} \mu_i^T \frac{\partial}{\partial\theta_i} \begin{pmatrix} \Sigma^{[s]}(\overset{\circ}{\theta}) \\ \begin{bmatrix} Z_{\overset{\circ}{\theta},0} & W_{\overset{\circ}{\theta},0} \end{bmatrix}^{[s]} \\ \begin{bmatrix} Z_{\overset{\circ}{\theta},1} & W_{\overset{\circ}{\theta},1} \end{bmatrix}^{[s]} \\ \vdots \end{pmatrix} = 0 \tag{3.34}$$

is true where $[Z_{\theta,i}, W_{\theta,i}]$ denotes the i-th Markov matrix of the system $[Z_\theta(z)W_\theta(z)]$.

Now the set of systems in $S(p,m,\delta)$ is parameterized in a one-to-one manner by the elements of the analytical manifold Θ. Hence, for all initial segments of the Markov matrix sequence of $[Z_\theta(z)W_\theta(z)]$ of length greater than the McMillan degree δ of $[Z_\theta(z)W_\theta(z)]$, taken together with $\Sigma(\theta)$, the Jacobian of the map from Θ to these quantities has full rank. But (3.34) cannot hold in that case, and so we conclude that $H(\overset{\circ}{\theta})$ has full rank as required.

Next we need

LEMMA 3.6. *Subject to the general hypotheses of the theorem and hypotheses INP A(1)-(4)*

$$\frac{\partial^2 L_N}{\partial\theta_i\partial\theta_j}(y_1^N;u,\theta) \to \frac{\partial^2 L_N}{\partial\theta_i\partial\theta_j}(\theta), \qquad 1 \le i,j \le \eta, \qquad \text{a.s.} \tag{3.35}$$

as $N \to \infty$ *for all* $\theta \in \Theta$, *uniformly over any compact set* $D \subset \Theta$.

<u>Conclusion of the proof of asymptotic normality under input hypotheses A.</u> The relation

$$-\frac{1}{\sqrt{N}} \frac{\partial L_N}{\partial\theta_i}(y_1^N;u,\overset{\circ}{\theta}) = \frac{1}{\sqrt{N}} \frac{\partial^2}{\partial\theta\partial\theta_i} L_N(y_1^N;u,\theta_{N,i})(\hat{\theta}_N - \overset{\circ}{\theta}) \qquad \forall N \in \mathbb{Z}_1 \tag{3.36}$$

implies that if $\frac{1}{\sqrt{N}} \frac{\partial L_N}{\partial\theta_i}(y_1^N;u,\overset{\circ}{\theta})$ converges in distribution then so does the right hand side of (3.36). By lemmas 3.5, 3.6 together with $\hat{\theta}_N \to \overset{\circ}{\theta}$ a.s. as $N \to \infty$, such an asymptotic distribution for the right hand side would be identical to that of the i-th entry of the η component vector $H(\overset{\circ}{\theta}) \frac{1}{\sqrt{N}} (\hat{\theta}_N - \overset{\circ}{\theta})$.

Hence, if the left hand side of (3.36) converged in distribution to

$$N\left(\overset{\circ}{\theta}, \begin{pmatrix} P(\overset{\circ}{\theta}) & Q(\overset{\circ}{\theta}) \\ Q^T(\overset{\circ}{\theta}) & R(\overset{\circ}{\theta}) \end{pmatrix}\right),$$

then

$$\frac{1}{\sqrt{N}}(\hat{\theta}_N - \overset{o}{\theta}) \xrightarrow{\text{dist}} N\left(0, H^{-1}(\overset{o}{\theta})\begin{bmatrix} P(\overset{o}{\theta}) & Q(\overset{o}{\theta}) \\ Q^T(\overset{o}{\theta}) & R(\overset{o}{\theta}) \end{bmatrix} H^{-1}(\overset{o}{\theta})\right)$$

as $N \to \infty$, which is our desired result.

The fact that the right hand side converges in distribution is stated in the following lemma:

LEMMA 3.7. *Subject to the general conditions of the theorem and hypotheses INP A(1-5). The sequence of random variables* $\{\sqrt{N}\,\frac{\partial L_N}{\partial \theta}(y^N; u, \overset{o}{\theta}); N \in \mathbb{Z}_1\}$ *converges in distribution* $N\left(0; \begin{bmatrix} P & Q \\ Q & R \end{bmatrix} \overset{o}{\theta}\right)$, *where* $Q = 0$ *and* P *and* R *are given in (3.37) and (3.38) below:*

The entries in the matrix $P(\overset{o}{\theta})$ *are given by*

$$P_{ij}(\overset{o}{\theta}) = E\left(\mathrm{Tr}[\Sigma^{-1}(\overset{o}{\theta})\frac{\partial \Sigma(\theta)}{\partial \theta_i}] + \mathrm{Tr}[(w_o w_o^T)\frac{\partial \Sigma^{-1}(\theta)}{\partial \theta_i}]\right)\Bigg|_{\theta=\overset{o}{\theta}}$$

$$\times \left(\mathrm{Tr}[\Sigma^{-1}(\overset{o}{\theta})\frac{\partial \Sigma(\theta)}{\partial \theta_j}] + \mathrm{Tr}[(w_o w_o^T)\frac{\partial \Sigma^{-1}(\theta)}{\partial \theta_j}]\right)\Bigg|_{\theta=\overset{o}{\theta}}$$

$$= 2\,\mathrm{Tr}[\Sigma^{-1}(\overset{o}{\theta})\frac{\partial \Sigma(\overset{o}{\theta})}{\partial \theta_i}\Sigma^{-1}(\overset{o}{\theta})\frac{\partial \Sigma(\overset{o}{\theta})}{\partial \theta_j}], \quad 1 \le i,j \le \frac{p(p+1)}{2} \tag{3.37}$$

(which is equivalent to the corresponding expression derived by Dunsmuir [1979]). The entries of the matrix R *are given by*

$$R_{ij} = \mathrm{Tr}\, M^{i,j}(\overset{o}{\theta})\Sigma(\overset{o}{\theta}) + \frac{1}{2\pi}\int_0^{2\pi}\mathrm{Tr}[W_{\overset{o}{\theta}}^{-1}(e^{i\lambda})\frac{\partial W_{\overset{o}{\theta}}}{\partial \theta_j}(e^{i\lambda})\Sigma^{-1}(\overset{o}{\theta})$$

$$\times \frac{\partial W_{\overset{o}{\theta}}^T}{\partial \theta_i}(e^{-i\lambda})W_{\overset{o}{\theta}}^T(e^{-i\lambda})\Sigma(\overset{o}{\theta})]d\lambda \quad , \quad p\frac{(p+1)}{2} + 1 \le i,j \le \eta \tag{3.38(i)}$$

where

$$M^{i,j}(\overset{o}{\theta}) \triangleq \lim_{N\to\infty}\frac{1}{N}\sum_{k=1}^{N}\eta_k^{(i)}(\overset{o}{\theta})\eta_k^{(j)}(\overset{o}{\theta}) \quad , \tag{3.38(ii)}$$

where $\eta_k^{(i)}(\theta)$ *denotes* $\sum_{j=0}^{\infty}\frac{\partial}{\partial \theta_i}K_j(\theta)u_{k-j}$, *when* $K_j(\theta)$ *is the coefficient matrix of* z^j *appearing in* $W_\theta^{-1}(z)\tilde{Z}_\theta(z)$ *(recall* $\nu^\infty(\theta) = W_\theta^{-1}(z)\tilde{Z}_\theta(z)u(z) + W_\theta^{-1}(z)W_{\overset{o}{\theta}}(z)W(z)$.

REFERENCES

Anderson, T. W. *The Statistical Analysis of Time Series.* John Wiley, New York, 1971.

Aström, K. J., T. Bohlin, and S. Wensmark. *Automatic Construction of Linear Stochastic Dynamic Models for Stationary Industrial Processes with Random Disturbances Using Operating Records.* IBM Nordic Laboratory, Sweden, TP 18.150 Technical Paper, June 1, 1965.

Box, G. E. P., and G. M. Jenkins. *Time Series Analysis-Forecasting and Control.* Holden-Day, San Francisco, 1970.

Byrnes, C. I. *The moduli space for linear dynamical systems,* Geometric Control Theory (C. Martin and R. Hermann, eds.), Math Sci Press, Brookline, MA, 1977, pp. 229-276.

Caines, P. E. *Linear Stochastic Systems.* John Wiley, New York, to appear.

Caines, P. E. *Prediction error identification methods for stationary stochastic processes,* IEEE Trans. Auto. Cont., Vol. AC-21 (1976), pp. 500-506.

Caines, P. E., and J. Rissanen. *Maximum likelihood estimation of parameters in multivariable Gaussian stochastic processes.* IEEE Trans. Information Theory, IT-20, no. 1 (1974), pp. 102-104.

Chan, S. W., G. C. Goodwin, and K. S. Sin. *Convergence and Properties of the Solutions of the Riccati Difference Equation.* Tech. Report 9201, Dept. of Electrical Engineering, University of Newcastle, Australia, 1982.

Clark, J. M. C. *The consistent selection of parameterizations in system identification.* Proc. JACC (1976), pp. 576-580.

Desoer, C. A. *Notes for a Second Course on Linear Systems.* Van Nostrand Rinehold, New York, 1970.

Deistler, M., W. Dunsmuir, and E. J. Hannan. *Vector linear time series models corrections and extensions,* Adv. Appl. Prob., Vol. 10 (1978), pp. 360-372.

Dunsmuir, W. *A central limit theorem for parameter estimation in stationary vector time series and its application to models for a signal observed with noise,* The Annals of Statistics, Vol. 7 (1979), no 3, pp. 490-506.

Dunsmuir, W., and J. Hannan. *Vector linear time series models,* Adv. Appl. Prob., Vol. 8 (1976), pp. 339-364.

Durbin, J. *Efficient estimation of parameters in moving-average model,* Biometrica, Vol. 46 (1959), pp. 306-316.

Gersch, W., and G. Kitagawa. *The prediction of time series with trends and seasonalities,* ASA/Census Time Series Research Report, December 1981.

Goodrich, R. L., and P. E. Caines (a). *Necessary and sufficient conditions for local second order identifiability,* IEEE Trans. on Auto. Contr., Vol. AC-24, no. 1, (February 1979), pp. 125-127.

Goodrich, R. L., and P. E. Caines (b). *Linear system identification from non-stationary cross-sectional data,* IEEE Trans. on Auto. Contr., Vol. AC-24, no. 3, (June 1979), pp. 403-411.

Hannan, E. J. *The statistical theory of linear systems,* Developments in Statistics (P. Krishnaiah, ed.). Academic Press, New York, 1979, pp. 83-121.

Hannan, E. J., W. Dunsmuir, and W. Deistler. *Estimation of vector ARMAX models,* J. of Multivariate Analysis, Vol. 10, no. 3 (September 1980), pp. 272-295.

Hazewinkel, M., and R. E. Kalman. *On invariants, canonical forms and moduli for linear, constant, finite dimensional dynamical systems,* Mathematical Systems Theory, Udine 1975. Springer Verlag, New York, 1976, pp. 48-60.

Hazewinkel, M. *Representations of quivers and moduli of linear dynamical systems,* The 1976 Ames Research Center Conference on Geometric Control Theory, Math Sci Press, Brookline, MA, 1977, pp. 277-289.

Heymann, M. *Structure and Realization Problems in the Theory of Dynamical Systems,* Springer CISM Courses and Lectures, No. 204, New York, 1975.

Kendall, M. G., and A. Stuart. *Advanced Theory of Statistics.* Griffin, London, 1948.

Kelley, J. L. *General Topology.* D. Van Nostrand, Princeton, New Jersey, 1964.

Ljung, L. *On the consistency of prediction error identification methods.* System Identification, Advances and Case Studies (R. K. Mehra, D. G. Lainiotis, eds.). Academic Press, New York, 1976, pp. 121-164. See also: Report 7405 Lund Institute of Technology, Dept. of Automatic Control, February 1974.

Ljung, L., and P. E. Caines. *Asymptrophic normality of prediction error estimators for approximate system models,* Stochastics, Vol. 3 (1979), pp. 29-46.

Mann, H. B., and A. Wald. *On the statistical treatment of linear stochastic difference equations,* Econometrica, Vol. 11, nos. 3 & 4 (1943), pp. 173-220.

Mayne, D. Q. *Parameter estimation,* Automatica, Vol. 3 (1966), pp. 245-255.

Mehra, R. K. *Identification of stocahstic linear dynamics systems using Kalman filter representation,* AIAA J., Vol. 9, no. 1 (January 1971), pp. 28-31.

Mehra, R. K. *On the identification of variances and adaptive Kalman filtering,* IEEE Trans. Aut. Contr., Vol. AC-15, no. 2 (April 1970), pp. 175-184.

Rissanen, J. *A fast algorithm for optimum linear predictors,* IEEE Trans. Aut. Contr., Vol. AC-18 (Oct. 1973), pp. 555.

Rissanen, J., and L. Barbosa. *Properties of infinite covariance matrices and stability of optimum predictors,* Information Sciences, Vol. 1 (1969), pp. 221-236.

Rissanen, J., and P. E. Caines. *The strong consistence of maximum likelihood estimators for ARMAX processes,* The Annals of Statistics, Vol. 7, no. 12 (1979), pp. 297-315.

Rissanen, J., and L. Ljung. *Estimation of optimum structures and parameters for linear systems,* Mathematical Systems Theory, Udine 1975, pp. 92-110.

Scweppe, F. C. *Uncertain Dynamic Systems.* Prentice Hall, Englewood Cliffs, New Jersey, 1973.

Wald, A. *Note on the consistency of the maximum likelihood estimate.* Ann. Math. Statist., Vol. 20 (1949), pp. 595-601.

Walker, A. M. *Large-sample estimation of parameters for moving-average models,* Biometrica, Vol. 48 (1961), pp. 343-357.

Whittle, P. *Estimation and information in stationary time series,* Ark. Mat., Vol. 2, no. 23 (1952), pp. 423-434.

MCGILL UNIVERSITY
MONTREAL, QUEBEC
CANADA

GEOMETRIC QUESTIONS IN SYSTEMS IDENTIFICATION

David F. Delchamps

1. INTRODUCTION. Our objective is to investigate certain fundamental geometric questions which arise in the analysis of the following identification problem:

<u>Given</u>: (1) Two vector-valued stochastic processes $y(t)$, $e(t)$, $t \in \mathbb{R}$ or $\mathbb{Z}$

(2) A set of Σ of finite-dimensional, time-invariant, continuous- or discrete-time linear systems serving as "candidate models" under the assumption that y and e are respectively the output and input of some (unknown) $\sigma_0 \in \Sigma$

and (3) Observations of $y(t)$ and/or $e(t)$ on some subset of $\mathbb{R}$ or $\mathbb{Z}$

<u>Determine</u>: some estimate of σ_0

During recent years, numerous approaches have been taken to this problem and its many variants; see, for example, [1], [8], [11], and [13]. In most treatments, it is desired to determine σ_0 recursively as the number of observations increases. The set Σ is parametrized in some manageable fashion, and objective functions $V^N(y,e,\sigma)$ (depending on $\sigma \in \Sigma$ and on the observations up to time N) are defined; the minimum of V^N over Σ occurs at the system which "best" accounts for the observations up to time N.

At this point, technical conditions are introduced which ensure that the minimizing values of σ converge to σ_0 as N becomes infinite. Generally included are probabilistic assumptions about y and e along with restrictions on the model set Σ. Often, these conditions enable one to show that, as N increases, V^N approaches a function $V:\Sigma \to \mathbb{R}$ whose unique minimum occurs at σ_0; in this case the problem of identifying σ_0 becomes the problem of minimizing the deterministic function V.

If the candidate model set Σ is a manifold, one might wish to minimize the function $V:\Sigma \to \mathbb{R}$ by following the flow induced on Σ by the gradient vector field $-\text{grad } V$ taken with respect to some Riemannian metric on Σ. Such an algorithm converges globally if and only if the "true" system σ_0 is the unique critical point of the function V and is also locally and globally attracting for the vector field $-\text{grad } V$. These are indeed strong requirements; it is easy to show (see [10]) that if on a manifold M there exists a vector field X with a single locally and globally attracting critical point, then M is contractible. A stronger statement is

MILNOR'S THEOREM. ([15], Theorem 2.2) *If there exists a vector field* X *on a manifold* M *with a single, locally and globally attracting critical point, then* M *is diffeomorphic to* $\mathbb{R}^n$, *where* $n = \dim(M)$.

Milnor's Theorem is evidence that the geometric properties of Σ have a great deal to do with the critical point behavior of functions $V:\Sigma \to \mathbb{R}$; of particular interest to us are those functions V which arise as objective functions in identification problems. More often than not, such functions satisfy the criterion that $V^{-1}(-\infty,a]$ be a compact subset of Σ for every $a \in \mathbb{R}$. Morse Theory provides the essential link between the topology of Σ and the critical point behavior of those functions $V:\Sigma \to \mathbb{R}$ which have compact sublevel sets. It is Morse Theory, in fact, on which the proof of Milnor's Theorem is based.

By way of Morse Theory, it is possible to reinterpret many statements regarding convergence properties of identification algorithms as assertions about the geometric structure of the candidate model set Σ. On the other hand, studying the topology of Σ provides partial answers to questions such as:

a) Given an objective function $V:\Sigma \to \mathbb{R}$, does V have a single globally attracting critical point? More generally, does there exist a $V:\Sigma \to \mathbb{R}$ with this property?

b) Are there "interesting" submanifolds $\Sigma' \subset \Sigma$ on which such a V exists?

The relevance of these questions to the system identification problem should be readily apparent.

2. MANIFOLDS OF LINEAR SYSTEMS. We begin with an example which illustrates some of the ideas outlined in §1. In [1], the authors consider the maximum likelihood estimation of the coefficients of the relatively prime polynomials C(z) and A(z) appearing in the stable, minimum phase spectral factor $\frac{C(z)}{A(z)}$ associated with a discrete-time scalar Gaussian process y having rational spectrum. This problem is tantamount to identifying the linear system whose output is $\{y_t\}$ and whose input is the innovations process $\{e_t = y_t - E(y_t|y_{t-1},y_{t-2},\ldots)\}$. Since y is Gaussian, the residual process is ergodic, and a suitably normalized version of the likelihood function V^N converges to the deterministic function

$$V(\hat{a},\hat{c}) = \frac{1}{4\pi i} \oint_{|z|=1} \frac{\hat{A}^*(z)C^*(z)\hat{A}^*(z^{-1})C^*(z^{-1})}{A^*(z)\hat{C}^*(z)A^*(z^{-1})\hat{C}^*(z^{-1})}\, dz$$

Taking $\deg(\hat{A}) = \hat{n}_a$, $\deg(\hat{C}) = \hat{n}_c$, it is clear that V is defined at least on the open subset H of $\mathbb{R}^{\hat{n}_a} \times \mathbb{R}^{\hat{n}_c}$ consisting of those vectors

$(\hat{a},\hat{c})$ for which the polynomials $\hat{A}(z)$ and $\hat{C}(z)$ with coefficients $\hat{a}$ and $\hat{c}$ respectively are both asymptotically stable; H is the set of stable, minimum phase transfer functions with degree at most $\hat{n}_a$.

It is shown in [1] that if $\hat{n}_a = \deg(A)$ and $\hat{n}_c = \deg(C)$, then the unique global minimum for $V: H \to \mathbb{R}$ occurs at $(a,c) \in H$. It has been demonstrated recently by Byrnes and Krishnaprasad that (a,c) is the unique critical point for the (complete) vector field $-\text{grad } V$, and is globally attracting. In view of Milnor's Theorem in §1, it may be concluded that H is diffeomorphic to a Euclidean space.

Next, consider the set $\text{Rat}(n;\mathbb{R})$ of *all* real scalar linear systems with McMillan degree n. $\text{Rat}(n;\mathbb{R})$ is an open submanifold of $\mathbb{R}^{2n}$ which has been studied extensively by Brockett and Krishnaprasad [2,3], Byrnes and Duncan [6], Segal [17], Glover [11], and others. It is shown in [2] that $\text{Rat}(n;\mathbb{R})$ has $(n+1)$ connected components; thus for connectivity reasons alone a globally convergent gradient algorithm of the type studied in [1] cannot exist on $\text{Rat}(n;\mathbb{R})$.

We are interested in the multivariable generalizations of Rat(n)--namely, the spaces $\Sigma^n_{m,p}(\mathbb{F})$ of m-input, p-output linear systems with McMillan degree n having coefficients in $\mathbb{F} = \mathbb{R}$ or $\mathbb{C}$. $\Sigma^n_{m,p}$ may be viewed either as a space of strictly proper rational matrices or as a space of $(p\times m)$-block Hankel matrices; it is a theorem of Clark (see [9] and also [12]) that $\Sigma^n_{m,p}(\mathbb{R})$ (respectively $\Sigma^n_{m,p}(\mathbb{C})$) is a real (resp. complex) analytic manifold of dimension $n(m+p)$. The manifolds $\Sigma^n_{m,p}(\mathbb{F})$ have been explored by Kalman and Hazewinkel [12], Byrnes and Duncan [5,7], Hermann and Martin [14], and others.

$\Sigma^n_{m,p}$ is clearly not compact; thus we cannot expect to learn a great deal about the topology of $\Sigma^n_{m,p}$ by examining the tangent bundle $T\Sigma^n_{m,p}$. Fortunately, linear system theory reveals a natural bundle over $\Sigma^n_{m,p}$. Let $\tilde{\Sigma}^n_{m,p}(\mathbb{F})$ be the open subset of $\mathbb{F}^{n^2} \times \mathbb{F}^{nm} \times \mathbb{F}^{pn}$ consisting of matrix triples (A,B,C) minimally realizing systems in $\Sigma^n_{m,p}(\mathbb{F})$; Byrnes and Hurt [7] along with Hazewinkel and Kalman [12] have shown that $\tilde{\Sigma}^n_{m,p}(\mathbb{F})$ is a real analytic $GL_n(\mathbb{F})$ principal bundle over $\Sigma^n_{m,p}(\mathbb{F})$. The projection $\pi: \tilde{\Sigma} \to \Sigma$ is the map

$$(A,B,C) \to C(Is-A)^{-1}B \quad .$$

Associated with $\tilde{\Sigma}^n_{m,p}(\mathbb{F})$ is the unique analytic vector bundle X over $\Sigma^n_{m,p}(\mathbb{F})$ for which $\tilde{\Sigma}^n_{m,p}(\mathbb{F})$ is the frame bundle. The fibre of X over $\sigma \in \Sigma^n_{m,p}$ may be viewed as the abstract state space of the system σ.

Perhaps the most obvious question one might ask is whether $\tilde{\Sigma} \to \Sigma$ and $X \to \Sigma$ are trivial bundles. Byrnes and Hurt [7] show that $\tilde{\Sigma}^n_{m,p} \to \Sigma^n_{m,p}$ is a trivial bundle if and only if $\min(m,p) = 1$; the same is therefore true for $X \to \Sigma^n_{m,p}$. This result precludes the existence of continuous canonical forms for systems which have more than one input and more than one output.

In what follows, we investigate other features of the bundles $\tilde{\Sigma}$ and X, and draw appropriate conclusions about the geometric structure of Σ. By way of Morse Theory, these results have implications with regard to the critical point behavior of objective functions on Σ.

3. METRICS ON X AND CONNECTIONS IN $\tilde{\Sigma}$. It follows from a partition of unity argument that every C^∞ real or complex vector bundle admits a C^∞ Riemannian or Hermitian metric. It is not in general true that a real analytic vector bundle admits a real analytic metric. In [10], it is shown that there exist three natural real analytic metrics on the state bundle $X \to \Sigma^n_{m,p}(\mathbb{F})$, $\mathbb{F} = \mathbb{R}$ or $\mathbb{C}$. To define an analytic metric on X, it suffices to find an $(n\times n)$ analytic positive definite symmetric (or Hermitian) matrix-valued function P on $\tilde{\Sigma}^n_{m,p}$ which transforms according to

$$P(g^{-1}Ag, g^{-1}B, Cg) = g^{-1}P(A,B,C)(g^{-1})^*$$

for $(A,B,C) \in \tilde{\Sigma}$ and $g \in GLn(\mathbb{F})$. The X-frames $(A,B,C) \in \tilde{\Sigma}$ which are orthonormal with respect to the metric are the solutions of

$$P(A,B,C) = I \quad .$$

One of the metrics defined in [10] is the *Riccati Metric*; the "P-matrix" for the Riccati Metric is P_*^{-1}, where P_* is the unique positive definite symmetric (or Hermitian) solution to the Algebraic Riccati Equation

$$P_*A + A^*P_* - P_*BB^*P_* + C^*C = 0 \quad .$$

Analyticity of the Riccati Metric is a consequence of the following technical result, which is of independent interest. (See [10], [4].)

LEMMA. *The function* $(A,B,C) \to P_*(A,B,C)$ *is analytic on* $\tilde{\Sigma}^n_{m,p}(\mathbb{F})$, $\mathbb{F} = \mathbb{R}$ or $\mathbb{C}$.

Recall that a connection in a $GLn(\mathbb{F})$ principal bundle $\tilde{M} \overset{\pi}{\to} M$ may be defined as the choice at each $u \in \tilde{M}$ of a dim(M)-dimensional subspace $H_u \subset T_u\tilde{M}$

(i) $u \to H_u$ is "smooth" over $\tilde{M}$
(ii) $\pi_*H_u = T_{\pi(u)}M \quad u \in \tilde{M}$
(iii) $Rg_*(H_u) = H_{ug} \quad u \in \tilde{M}, \ g \in GLn(\mathbb{F})$.

Here, $Rg: \tilde{M} \to \tilde{M}$ denotes the map $u \to ug$. For further details see [16] and [10]. Condition (ii) prompts the name "horizontal" for the subspaces H_u. The points $u \in \tilde{M}$ which are orthonormal frames with respect to a given metric on the associated vector bundle $X \overset{\rho}{\to} M$ form a regular submanifold $O\tilde{M} \subset \tilde{M}$. A connection $\{u \to H_u\}$ in $\tilde{M}$ is *compatible* with the metric if and only if $H_u \subset T_u(O\tilde{M})$ at each $u \in O\tilde{M}$.

From each metric on the *tangent bundle* of a manifold there arises a canonical compatible connection--the Levi-Civita connection--in the frame bundle of the tangent bundle. There is no such canonical connection compatible with a given metric on an arbitrary real vector bundle. In the case of the bundle $X \to \Sigma^n_{m,p}(\mathbb{F})$, however, there exist three analytic connections in $\tilde{\Sigma}^n_{m,p}(\mathbb{F})$ each of which is compatible with one of the aforementioned analytic metrics on X. In the case of the Riccati Metric, the horizontal subspace H_u at $u \in \tilde{\Sigma}$ is the kernel at u of the gln-valued one-form

$$P_*^{-1}\int_0^\infty e^{(A^*-P_*BB^*)t}[P_*dA-P_*dBB^*P_* + C^*dC]e^{(A-BB^*P_*)t}dt$$

In [10], connections in $\tilde{\Sigma}^n_{m,p}(\mathbb{F})$ play a central role in proving the following result.

THEOREM 1. *Let* $\mathbb{F} = \mathbb{R}$ or $\mathbb{C}$. *If* $\min(m,p) > 1$, *there does not exist a function* $V: \Sigma^n_{m,p}(\mathbb{F}) \to \mathbb{R}$ *with a single critical point which is locally and globally attracting for the vector field* $-\text{grad } V$.

Proof. (outline). The existence of such a V would required that $\Sigma^n_{m,p}(\mathbb{F})$ be contractible; see §1. We show that $\Sigma^n_{m,p}(\mathbb{R})$ is not contractible when $\min(m,p) > 1$ by exhibiting a loop γ based at $\sigma \in \Sigma^n_{m,p}(\mathbb{R})$ whose horizontal lift $\tilde{\gamma}$ (see [16]) with respect to one of our connections links "opposite sides" of the fibre $GLn(\mathbb{R}) = \pi^{-1}(\sigma) \subset \tilde{\Sigma}^n_{m,p}(\mathbb{R})$. This loop is homotopically nontrivial, since it gives rise to a nontrivial element in the reduced holonomy group [16] of $\tilde{\Sigma}^n_{m,p}(\mathbb{R})$.

Since $\tilde{\Sigma}^n_{m,p}(\mathbb{C})$ is simply connected (see [10], [6]) we bypass the fundamental group and prove that $\pi_2(\Sigma^n_{m,p}(\mathbb{C})) \neq 0$ when $\min(m,p) > 1$ by exhibiting a copy of S^2 contained in Σ over which the trace of the curvature form (see [16] or [10]) of one of the connections has a nonzero integral. As it happens, this two-sphere sits under an embedded $S^3 \subset \tilde{\Sigma}$, and the map $\pi: S^3 \to S^2$ is the Hopf map.

The assumption $\min(m,p) > 1$ is essential; if $\min(m,p) = 1$, then the bundle $\tilde{\Sigma}^n_{m,p} \to \Sigma^n_{m,p}$ is trivial (c.f. §2) and it follows in the real case that "opposite sides" of the same fibre lie in separate arc components of $\tilde{\Sigma}$. We also note that the vector field "-grad V" is not well-defined unless a Riemannian or Hermitian metric is specified on $\Sigma^n_{m,p}$; the property of being a global sink for -grad V is, however, independent of the metric.

4. AN APPLICATION OF MORSE THEORY. Morse Theory is the study of relationships between the topology of a manifold M and the critical point behavior of functions $V: M \to \mathbb{R}$ which have compact sublevel sets. The central results

of Morse Theory are the Morse Inequalities, of which we state a restricted version here (see [15]).

THE (WEAK) MORSE INEQUALITIES: *Let* $V : M \to \mathbb{R}$ *have compact sublevel sets and only nondegenerate critical points. Then if* $\mathbb{F}$ *is any field, we have*

$$c_i \geq \beta_i(M;\mathbb{F})$$

where c_i *denotes the number of critical points of index* i *and* $\beta i(M;\mathbb{F})$ *denotes the* i-th *Betti number of* M *with coefficients in* $\mathbb{F}$.

Recall that the *index* of a critical point p of a function $V : M \to \mathbb{R}$ is the number of negative eigenvalues of the Hessian of V at p; p is nondegenerate when the Hessian at p is nonsingular. A nondegenerate critical point of index zero is a local minimum.

The Morse Inequalities work in two directions: they yield quantitative information about critical point behavior given topological data; conversely, critical point behavior yields quantitative information about topology. We sketch here two examples which illustrate the applicability of Morse Theory to the problem at hand.

Consider first the homotopy exact sequence (see [18]) of the principal bundle $\tilde{\Sigma}^n_{m,p}(\mathbb{C}) \to \Sigma^n_{m,p}(\mathbb{C})$. At the end of the sequence, we have:

$$\to \pi_2(\tilde{\Sigma}) \to \pi_2(\Sigma) \to \pi_1(GLn(\mathbb{C})) \to$$

$$\to \pi_1(\tilde{\Sigma}) \to \pi_1(\Sigma) \to \pi_0(GLn(\mathbb{C})) = 0 \quad .$$

It is shown in [10] that $\pi_i(\tilde{\Sigma}^n_{m,p}(\mathbb{C})) = 0$ for $i < 2(\min(m,p)) - 1$. If we let $n = 1$ and $\min(m,p) > 1$, the above diagram gives

$$\pi_2(\Sigma) \simeq \pi_1(GL_1(\mathbb{C})) \simeq \mathbb{Z}$$

$$\pi_1(\Sigma) = 0 \quad .$$

We conclude from the Hurewicz Theorem that $H_2(\Sigma;\mathbb{Z}) \simeq \mathbb{Z}$. Thus for any field $\mathbb{F}$:

$$\beta_2(\Sigma^1_{m,p}(\mathbb{C});\mathbb{F}) = 1 \quad .$$

The Morse Inequalities immediately yield the following sharpened version of the complex part of Theorem 1.

THEOREM 2. *Let* $\min(m,p) > 1$. *Any objective function* $V : \Sigma^1_{m,p}(\mathbb{C}) \to \mathbb{R}$ *with compact sublevel sets and only nondegenerate critical points has at least one critical point of index two.*

In [10], there is defined on $\Sigma^1_{m,p}(\mathbb{F})$ a function V satisfying the hypotheses of the Morse Inequalities. In both the real and complex cases, V has exactly 2 min(m,p) critical points provided min(m,p) > 1. On $\Sigma^1_{2,2}(\mathbb{R})$, the critical points are distributed over indices according to $c_0 = c_2 = 1$, $c_1 = 2$; in the case of $\Sigma^1_{2,2}(\mathbb{C})$, we have $c_0 = c_2 = c_3 = c_5 = 1$.

Using the Morse Inequalities, we obtain strong constraints on the Betti numbers of $\Sigma^1_{m,p}(\mathbb{F})$. In [10] it is shown under the assumption min(m,p) > 1 that $V : \Sigma^1_{m,p}(\mathbb{C}) \to \mathbb{R}$ is always a "perfect Morse function" in the sense that the Morse Inequalities are always equalities; thus Morse Theory enables us to determine the homology of $\Sigma^1_{m,p}(\mathbb{C})$ completely when min(m,p) > 1.

We close with a quantitative result concerning the critical point behavior of objective functions on $\Sigma^n_{m,p}(\mathbb{F})$.

THEOREM. (see [10]) *Let* $\mathbb{F}$ *be* $\mathbb{R}$ *or* $\mathbb{C}$. *Any function* $V : \Sigma^n_{m,p}(\mathbb{F}) \to \mathbb{R}$ *with compact sublevel sets has at least* min(m,p) *critical points.*

The proof is based on the analysis of a bundle embedding $\psi : \tilde{\Sigma}^1_{m,p}(\mathbb{F}) \to \Sigma^n_{m,p}(\mathbb{F})$ for $n > 1$; using the data obtained from the study of the Morse function V above, it is possible to prove that certain characteristic classes of the bundles $\tilde{\Sigma}^n_{m,p}(\mathbb{F})$ do not vanish. Theorem 3 then follows from some well-known results in algebraic topology which link the characteristic classes of $\tilde{\Sigma}$ to the Lusternik-Schnirelmann category of Σ. Observe that Theorem 3 does not require that the critical points of V be nondegenerate.

ACKNOWLEDGEMENTS. The author wishes to acknowledge many helpful conversations with C. I. Byrnes, under whose supervision this research was conducted. Special thanks are also due to Peter Caines and Robert Hermann for organizing this exceptionally interesting conference.

REFERENCES

[1] Åström, K., and T., Söderström. *Uniqueness of the maximum likelihood estimates of the parameters of an ARMA model,* IEEE Trans. Aut. Contr., AC19 (1974), pp. 769-773.

[2] Brockett, R. W. *Some geometric questions in the theory of linear systems,* IEEE Trans. Aut. Contr., AC-21 (1976), pp. 449-455.

[3] Brockett, R. W., and P. S. Krishnaprasad. *A scaling theory for linear systems,* IEEE Trans. Aut. Contr., AC-25 (1980), pp. 197-207.

[4] Byrnes, C. I. *On the stabilizability of linear control systems depending on parameters,* Proceedings of the 18th IEEE Conf. on Decision and Control, Ft. Lauderdate, Florida, Dec. 1979, pp. 233-236.

[5] Byrnes, C. I. *The moduli space for linear dynamical systems,* Geometric Control Theory, (C. Martin and R. Hermann, eds.), Math Sci Press, Brookline, MA, 1977, pp. 229-276.

[6] Byrnes, C. I., and T. Duncan. *On certain topological invariants arising in system theory,* New Directions in Applied Mathematics (P. Hilton and G. Young, eds.), Springer-Verlag, New York, 1981, pp. 29-71.

[7] Byrnes, C. I., and N. Hurt. *On the moduli of linear dynamical systems,* Adv. Math. Studies in Anal., Vol. 4 (1979), pp. 83-122.

[8] Caines, P. E. *Prediction error identification methods for stationary stochastic processes,* IEEE Trans. Aut. Contr., AC-21 (1976), pp. 500-505.

[9] Clark, J. M. C. *The consistent selection of local coordinates in linear system identification,* Proceedings of the Joint Automatic Control Conference (1976), pp. 576-580.

[10] Delchamps, D. F. *The Geometry of Spaces of Linear Systems with an Application to the Identification Problem,* Ph.D. thesis, Harvard University, Cambridge, MA, 1982.

[11] Glover, K. *Structural Aspects of System Identification,* Ph.D. thesis, Massachusetts Institute of Technology, Cambridge, MA, Report No. ESL-R-516.

[12] Hazewinkel, M., and R. E. Kalman. *On invariants, canonical forms, and moduli for linear dynamical systems,* Lecture Notes on Economic-Mathematical System Theory Volume 131, Springer-Verlag, New York, 1976, pp. 48-60.

[13] Hazewinkel, M., and J. C. Willems, eds. *Stochastic Systems: The Mathematics of Filtering and Identification and Applications,* D. Reidel, Dordrecht, Holland, 1981.

[14] Hermann, R., and C. F. Martin. *Applications of algebraic geometry to systems theory: The McMillan degree and Kronecker indices of transfer functions as topological and holomorphic system invariants,* SIAM J. Contr., Vol. 16 (1978), pp. 743-755.

[15] Milnor, J. W. *Morse Theory,* Princeton University Press, Princeton, 1963.

[16] Nomizu, K. *Lie Groups and Differential Geometry,* Math. Soc. of Japan, Tokyo, 1956.

[17] Segal, G. *The topology of spaces of rational functions,* Acta Mathematica, Vol. 143 (1979), pp. 39-72.

[18] Steenrod, N. *The Topology of Fibre Bundles,* Princeton University Press, Princeton, 1951.

[19] Wilson, F. W. *The structure of the level surfaces of a Lyapunov function,* J. Diff. Eq., Vol. 3 (1967), pp. 323-329.

DEPARTMENT OF ELECTRICAL ENGINEERING
CORNELL UNIVERSITY
ITHACA, NEW YORK

SUFFICIENCY IN AUTOREGRESSIVE MODELS

Bradley W. Dickinson

1. INTRODUCTION. We consider the class of all stationary Gaussian, scalar, "lumped" time series (stochastic processes). A natural issue to explore when considering statistical inference problems involving members of this class is the existence of a nontrivial sufficient statistic. In particular, for the sake of "data compression," we might ask for the subclass which admits a sufficient statistic whose (Euclidean) dimension is independent of the number of observations made of the process. Arato [1], showed that this condition is satisifed only when the time series is an autoregression (AR) process. The intuitive explanation of this result is the following: the innovations representation of an AR process is a finite-impulse-response (FIR) or moving average filter which becomes time-invariant after a finite number of steps.

Since a sufficient statistic is a vector-valued function defined on the sample space, R^N, its properties may be investigated by geometric methods. There are very general results, due to Barankin and Katz [2,3], about the local minimal dimensionality of a sufficient statistic. Since the joint density function of a sample sequence from an AR process belongs to the exponential family, it is easy to separate the influence of the parameters and of the observations in the determination of the local minimal dimension [4]. However, this approach does not lead to the global results which would be desirable for applications: we would prefer to have a "universal data compression" function.

Our results are based on structural properties of low displacement rank matrices [5,6] and illustrate the essential nonlinear character of the appropriate representation of the log-likelihood function of an AR process. In a previous paper [7], we have discussed efficiencies of implementation that result from exploiting the structure of the sufficient statistic in particular inference problems. One might speculate about the need for a real complexity-based theory to incorporate order selection and precision issues in the spirit of Rissanen's work [8].

In a brief closing section, we point out the duality between the minimal dimension parameter space (identifiability) and minimal dimension sufficient statistic concepts as noted by Barankin [9]. For exponential families, the duality between sufficiency and identifiability was formally investigated by Picci [10], but without geometric considerations. In view of more recent work on canonical forms and identifiability for multivariable linear systems, it follows that local results are the best available in the general case.

2. SUFFICIENT STATISTICS OF VARIOUS SORTS FOR AR PROCESSES. Let $x = (x_1, x_2, \ldots, x_N)'$ be a set of observations from a p-th order stationary Gaussian AR process defined by the equation

$$x_i + a_1 x_{i-1} + \ldots + a_p x_{i-p} = \varepsilon_i \quad .$$

The $\{\varepsilon_i\}$ are *i.i.d.* Gaussian random variables with zero means and variance σ^2. We will consider a family of such processes whose parameters form an open subset of R^{p+1}. The family of density functions for x is described by the parametric density function

$$p(x; a_1, a_2, \ldots, a_p, \sigma^2) = p(x|\theta) = (2\pi)^{-N/2}(\det R_N)^{-1/2}\exp(-x'R_N^{-1}x/2)$$

where R_N is the Toeplitz covariance matrix

$$R_N = Exx' \quad .$$

After some simplification [11, pp. 274-276], we collect all the data dependent terms and obtain

$$\log p(x|\theta) = b_0(\theta) - a'S_p a/2\sigma^2$$

where $b_0(\theta)$ is a function depending on $a_1, \ldots, a_p$ and σ^2 but not $x, a' = (1, a_1, \ldots, a_p)$, and S_p is a $p+1$ by $p+1$ matrix given by

$$(S_p)_{ij} = \sum_{k=1}^{N-i-j} x_{i+k}x_{j+k} \ , \quad 0 \le i,j \le p \quad .$$

From the Neyman factorization theorem, the $(p+2)(p+1)/2$ elements in the upper triangular part of S_p form a sufficient statistic. If we write

$$\log p(x|\theta) = b_0(\theta) - \sum_{i=0}^{p}\sum_{j=0}^{i} b_{ij}(\theta)(S_p)ij$$

for appropriately defined function $b_{ij}(\theta)$, it is clear that the density function belongs to the exponential family. There are no linear dependence relations among the set of functions $\{1, b_{ij}(\theta), o \le j \le i \le p\}$, so the representation is in reduced form [4]; i.e., the number of terms in the sum is minimal.

With a sufficient statistic which grows in dimension with the square of the number of parameters, one immediately looks for possible redundancy. To my knowledge, this was first addressed in [7], perhaps because Arato's paper [1] was not very well known. In addition, the two lowest order examples ($p = 1$ and $p = 2$) are provably minimal dimension sufficient statistics; it takes a third point to see the difference between a linear and a quadratic function!

We may apply Barankin's solution for exponential families [4] and verify that redundancy is present. The minimal dimension sufficient statistic, in the neighborhood of a (regular) point x^*, has dimension given by the rank of the Jacobian matrix

$$\left(\frac{\partial (S_p)ij}{\partial x_k}\right)_{x=x^*} .$$

(This is a $(p+2)(p+1)/2$ by N matrix.) By regular point, we mean the rank remains constant over some neighborhood of the point. We will simplify this expression before attempting any calculation.

The key to simplification is the low displacement rank structure [5,6] of the matrix S_p. This is reflected in the following relation, holding for

$$1 \leq i,j \leq p : (S_p)_{ij} = (S_p)_{i-1,j-1} - x_i x_j - x_{N+1-i} x_{N+1-j} .$$

So S_p is determined, for example, by its first row $(S_p)_{0j}$, $0 \leq j \leq p$, and the end segments of data $(x_1,\ldots,x_p)$ and $(x_{N-p+1},\ldots,x_N)$. Here we have a sufficient statistic of dimension $3p + 1$! While this statistic is nice because only $p+1$ quantities need to be computed, we have lost functional minimality (necessity of the statistic). It is not, even locally, a function of the original sufficient statistic. For example, the sign of x_1 or x_N cannot be obtained.

Of course the problem is best viewed geometrically: the "real" statistic available from the decomposition of S_p is its first row together with the rank 2 symmetric, nonnegative definite residual matrix $[x_i x_j + x_{N+1-i} x_{N+1-j}]$. This may be locally parametrized with $2p - 1$ numbers, giving the minimal dimension, $3p$. Since we are free to discard subsets of the observation space of zero measure, we obtain an "almost sure" global statistic by using the first two columns of this matrix, omitting the first entry of the second column. Another, more analytically convenient choice is to take the first two columns of the Cholesky factorization of the matrix; these will almost surely be the two nonzero columns since the matrix has rank two. We denote these two columns by $(c_{11},c_{12},\ldots,c_{1p})' = c_1$ and $(0,c_{22},\ldots,c_{2p})' = c_2$. Also let Σ_p be the $p+1$ by $p+1$ symmetric Toeplitz matrix whose first row is the same as the first row of S_p.

Now we are able to present an alternative form for the log-likelihood function. Using the low displacement rank property of S_p together with the factored form of the low rank residual matrix, we have

$$S_p = \Sigma_p - L_1L_1' - L_2L_2'$$

where L_1 and L_2 are strictly lower triangular Toeplitz matrices whose first columns are c_1 and c_2 respectively [5] (with a zero appended for the first entry of each). Let us enumerate the components of the sufficient statistic as $t_1(x) = (S_p)_{oo},\ldots,t_{p+1}(x) = (S_p)_{op}$; $t_{p+2}(x) = c_{11},\ldots,t_{2p+1}(x) = c_{1p}$, $t_{2p+2}(x) = c_{22},\ldots,t_{3p}(x) = c_{2p}$. Then for suitable functions $\beta_k(\theta)$, $1 \leq k \leq 3p$, depending on $a_1,a_2,\ldots,a_p$, and σ^2, we have

$$\log p(x|\theta) = b_0(\theta) - \sum_{k=1}^{p+1} \beta_k(\theta)t_k(x) + \left(\sum_{k=p+2}^{2p+1} \beta_k(\theta)t_k(x)\right)^2$$

$$+ \left(\sum_{k=2p+2}^{3p} \beta_k(\theta)t_k(x)\right)^2$$

This quadratic representation, unlike the linear one, captures the structure (almost surely) of the sufficient statistic, vis-a-vis minimal dimension and warrants further investigation. We note with interest that some geometrical analysis of exponential families has appeared in work by Efron [12].

We would also like to point out that the stationarity assumption can be relaxed somewhat. We require the AR model to hold but allow the initial covariance of the state of the model to enter as additional parameters. The details have been worked out by Jeff Klein.

3. DUALITY WITH IDENTIFIABILITY. In this section we are not restricting our consideration to AR models; we assume that we have a parametric family $\{p(x|\theta)\}$ with mild regularity properties. The general theory of minimal dimensionality of a sufficient statistic [2,3] involves the rank of a matrix of mixed second partial derivatives of $\log p(x|\theta)$ with respect to the observations, x_i, and the parameters.

By introducing a prior probability on the parameters, with density $q(\theta)$, we obtain $p(x,\theta) = p(x|\theta)q(\theta)$, the joint density of x and θ. We may integrate to obtain the marginal density for x, say $f(x)$, and divide to obtain $p(\theta|x) = p(x|\theta)q(\theta)/f(x)$. But now we notice that the mixed second partial derivatives of $\log p(\theta|x)$ are the same as those of $\log p(x|\theta)$, regardless of the choice of $q(\theta)$. We have simply reversed the roles of observations and parameters and, as discussed by Barankin [4], the results for minimal dimension sufficient statistics give results for minimal dimension parametrizations, i.e., identifiability.

A more algebraic approach to duality was presented by Picci [10]. He did not concern himself with the geometric structure underlying particular examples of the exponential family such as the AR processes. However, he provided a very nice axiomatic approach to sufficiency and identifiability based on canonical factorization of mappings--very suggestive of realization theory for linear systems. We hope to continue our exploration of this area by examining the role of such quadratic expressions for the log-likelihood as developed above.

ACKNOWLEDGMENT. For help in discovering the statistical literature, I want to thank D. Thomson and G. Terdik. This work was supported in part by NSF Grant ENG77-28523.

REFERENCES

1. M. Arato. *On the sufficient statistics for stationary Gaussian random processes,* Theory Prob. Appl., Vol. 6, no. 2 (1961), pp. 199-201.

2. E. W. Barankin and M. Katz, Jr. *Sufficient statistics of minimal dimension,* Sankhya, Vol. 21 (1959), pp. 217-246.

3. E. W. Barankin and M. Katz, Jr. *A note on functional minimality of sufficient statistics,* Sankhya, Vol. 23 (1961), pp. 401-404.

4. E. W. Barankin. *Application to exponential families of the solution to the minimal dimensionality problem for sufficient statistics,* Ball. Inst. Internat. Stat., Vol. 38 (1961), pp. 141-150.

5. T. Kailath, S. Kung, and M. Morf. *Displacement ranks of matrices and linear equations,* J. Math. Analy. Appl., Vol. 68 (1979), pp. 395-407.

6. B. Friedlander, M. Morf, T. Kailath, and L. Ljung. *New inversion formulas for matrices classified in terms of their distance from Toeplitz matrices,* Lin. Alg. Appl., Vol. 27 (1979), pp. 31-60.

7. B. W. Dickinson. *Properties and application of Gaussian autoregressive processes in detection theory,* IEEE Trans. Inform. Theory, Vol. It-27 (1981), pp. 343-347.

8. J. Rissanen. *Universal prior for parameters and estimation by minimum description length,* miuate communication.

9. E. W. Barankin. *Sufficient parameters: solution of the minimal dimensionality problem,* Ann. Inst. Statist. Math., Vol. 12 (1960), pp. 91-118.

10. G. Picci. *Some connections between the theory of sufficient statistics and the identifiability problem,* SIAM J. Appl. Math., Vol. 33 (1977), pp. 383-398.

11. G. E. P. Box and G. M. Jenkins. *Time Series Analysis: Forecasting and Control,* rev. ed., Holden Day, San Francisco, 1976.

12. B. Efron. *The geometry of exponential families,* Ann. Stat., Vol. 6, no. 2 (1978), pp. 362-376.

DEPARTMENT OF ELECTRICAL ENGINEERING AND COMPUTER SCIENCE
PRINCETON UNIVERSITY
PRINCETON, NEW JERSEY
U.S.A.

THE STRUCTURE OF ARMA SYSTEMS IN RELATION TO ESTIMATION

Manfred Deistler

1. INTRODUCTION. In this contribution we shall be concerned with some of the properties of the structure of ARMA systems which are relevant for estimation.

ARMA systems are of the form

$$\sum_{i=0}^{p} A(i)y(t-i) = \sum_{i=1}^{q} B(i)\varepsilon(t-i) \qquad (1.1)$$

where $A(i), B(i) \in \mathbb{R}^{s\times s}$ are parameter-matrices, where $y(t)$ are the (observed) outputs and where $\varepsilon(t)$ are (nonobserved) white noise random inputs, i.e.

$$E\varepsilon(t) = 0 \quad ; \quad E\varepsilon(s)\varepsilon'(t) = \delta_{st}\cdot\Sigma \qquad (1.2)$$

Let

$$a(z) = \sum_{i=0}^{p} A(i)z^{i} \quad ; \quad b(z) = \sum_{i=0}^{q} B(i)z^{i} \quad .$$

The assumption

$$\det a(z) \not\equiv 0 \qquad (1.3)$$

is part of the definition of an ARMA system. Adding a term $\Sigma D(i)x(t-i)$ on the right hand side of (1.1), where $x(t)$ are observed inputs, gives an ARMAX system. In this more general case results analogous to those presented here hold and we restrict ourselves to the ARMA case only to keep notation simple.

Every linear, time-invariant, finite dimensional, discrete-time system with white noise inputs can be written in ARMA form. Every ARMA system can be transformed into state space form.

A linearly regular, (wide sense) stationary solution of an ARMA system is called an ARMA process.

Throughout the paper we will make the following assumptions: The transfer functions

$$k(z) = a^{-1}(z)\cdot b(z) \qquad (1.4)$$

are causal, i.e. they have a convergent power series expansion in a neighbourhood about zero:

$$k(z) = \sum_{i=0}^{\infty} K(i)z^i \quad . \tag{1.5}$$

Furthermore we assume

$$\Sigma > 0 \tag{1.6}$$

and

$$K(0) = I \quad . \tag{1.7}$$

The minimum phase assumption

$$\det b(z) \neq 0 \qquad |z| < 1 \tag{1.8}$$

will not be imposed explicitly, unless mentioned otherwise, mainly because this assumption is needed only due to the fact that $\varepsilon(t)$ cannot be observed, but also because the consequences of this assumption do not essentially change our results, but complicate the statements. For similar reasons also the stability assumption

$$\det a(z) \neq 0 \qquad |z| \leq 1$$

will not be imposed explicitly unless mentioned otherwise.

The transfer function k and Σ can be obtained from $(y(t))$. On the other hand, the parameters of interest for the description of an ARMA system are $(A(o),\ldots,A(p),B(o),\ldots,B(q),\Sigma)$ and thus the relation between k and (a,b) is of great interest.

A pair (a,b) of polynomial matrices (obeying our assumptions) satisfying (1.4) is called a left matrix fraction description (MFD) of the transfer function k. The class of all ARMA systems (for fixed s) can be described (omitting Σ) via its parameter-matrices $(A(o),A(1),\ldots,B(o),B(1),\ldots)$ as a subset, A say, of $\mathbb{R}^{\mathbb{N}}$. We identify (a,b) with the corresponding parameter matrices. Let π denote the mapping attaching to every $(a,b) \in A$ the corresponding transfer function $a^{-1} \cdot b$ and let U_A denote the image of A by π. The equivalence kernel of π is called the relation of *observational equivalence*. (If $S \subset A$, then) $\pi^{-1}(k)$ $(\pi^{-1}(k) \cap S)$ is called the k-*equivalence class* (in S) or the class of *observationally equivalent* MFD's (in S).

For estimation, parameter-spaces which are subsets of Euclidean spaces (rather than subsets of infinite dimensional spaces) are more convenient. Thus we have to break the set of all ARMA systems into parts. Also for estimation, we want unique descriptions of the transfer functions by MFD's. A subclass of ARMA systems (a subset $S \subset A$) is called *identifiable*, if π restricted to S

is injective, i.e. if within this class (a,b,Σ) are uniquely determined from (k,Σ).

To avoid redundancy we often restrict ourselves to (relatively) left prime MFD's (a,b), i.e. to MFD's whose (polynomial matrix) greatest common left divisors are unimodular (a polynomial matrix u is called unimodular, if $\det u = \text{const} \neq 0$). The classes of observational equivalence are characterized by the fact, that if (a,b) is (relatively) left prime then $(\bar{a},\bar{b})$ is observationally equivalent to (a,b) if and only if there exists a polynomial matrix u such that

$$(\bar{a},\bar{b}) = u(a,b) \tag{1.10}$$

if $(\bar{a},\bar{b})$ is left prime too, then u in (1.10) is unimodular [11].

2. THE PARAMETRIZATION OF ARMA SYSTEMS. There are several ways to break the set of all ARMA systems into parts, to obtain finite dimensional parameter-spaces and to obtain identifiability.

In many applications, mainly in the ARMAX case, the parameters have direct physical interpretation and there is additional a-priori knowledge available in the form of a-priori restrictions on the parameters; e.g. the maximum lag-length for every variable may be a-priori known. Then the question arises whether these a-priori restrictions can guarantee identifiability. In this case the word *structural identifiability* is used in econometrics.

If there are no additional a-priori restrictions on the ARMA parameters, we are free to pick out a suitable representative from every equivalence class. Let $S \subset A$; a mapping $f : S \to S$ attaching to every element from S a representative of the corresponding equivalence class is called a *canonical form* (for S). We use the same word also for the values of the function f, i.e. for the prescribed representatives. If $A = S$ then we will speak about canonical forms without mentioning the reference set A.

A third approach, also for the case of no additional a-priori restrictions is the "overlapping" description of the manifold $M(\nu)$, explained later in this paper.

As already mentioned, the a-priori information available may be in the form of a prescription of the maximum lags for every variable, i.e. the degrees of every column of (a,b) are prescribed a-priori. For the sake of simplicity in this case we assume

$$A(o) = I \quad (\text{and thus } B(o) = I) \tag{2.1}$$

and that the other parameters are "freely varying," i.e. that there are no additional "overidentifying" restrictions. It should be noted however that the case of overidentifying restrictions is of great practical importance since the over-identifying restrictions may reduce the dimension of the parameter-space considerably and thus increase the number of "degrees of freedom" for estimation.

Let $\alpha = (p_1,\dots,p_s,q_1,\dots,q_s)$ denote the specified (maximal) degrees of the columns $a_1\dots a_s, b_1\dots b_s$ of (a,b) and let $a_i(j)$ and $b_i(j)$ denote the coefficients of z^j in a_i and b_i respectively. Every MFD (a,b) with specified degrees α can be identified with

$$\theta = \text{vec}(a_1(1)\dots a_1(p_1), a_2(1)\dots a_s(p_s), b_1(1)\dots b_s(p_s)) \in \mathbb{R}^n \quad ;$$

$$n = s(p_1 + p_s + \dots + q_s) \quad .$$

Let

$$C_\alpha = (a_1(p_1)\dots a_s(p_s) b_1(p_1)\dots b_s(p_s))$$

denote the column-end-matrix of (a,b). If we assume that the column-end-matrix of (a,b) has rank s, then the polynomial matrix u in (1.10) must be a constant. Thus the set, $\theta_\alpha^{(1)} \subset \mathbb{R}^n$ say, of all MFD's with lag-lengths prescribed by α, which are left prime, where the column-end-matrix C_α has rank s and where (2.1) holds, is identifiable [12]. For more general conditions for structural identifiability see [12][4]. By $U_\alpha^{(1)}$ we denote the set of all transfer functions corresponding to $\theta_\alpha^{(1)}$.

The usual canonical forms start from a transfer function

$$\tilde{k}(z) = \sum_{i=1}^{\infty} K(i) z^{-i} \tag{2.2}$$

rather than from $k(z)$, since $\tilde{k}(z)$ has the advantage of being strictly proper (under our causality assumption), i.e.

$$\lim_{z\to\infty} \tilde{k}(z) = 0$$

and therefore for every MFD

$$\tilde{k}(z) = \tilde{a}^{-1}(z) \cdot \tilde{b}(z) \tag{2.3}$$

the degree of the i-th row of $\tilde{a}$, μ_i say, is greater than the degree of the corresponding row of $\tilde{b}$. If $(\tilde{a},\tilde{b})$ is a canonical form for $\tilde{k}$ then the corresponding canonical form for k is given by

$$(a(z),b(z)) = \{\text{diag}\, z^{\mu_i}\}\ (\tilde{a}(z^{-1}), \tilde{a}(z^{-1}) + \tilde{b}(z^{-1})) \quad . \tag{2.4}$$

A common canonical form, called *Echelon form* (see [10][20]) can be obtained as follows: From (2.3) we have:

$$0 = (\tilde{A}(o),\dots\tilde{A}(p))H_{p+1} \tag{2.5}$$

where

$$H_{p+1} = \begin{bmatrix} K(1),K(2),K(3),\ldots\ldots\ldots \\ K(2),K(3),K(4),\ldots\ldots\ldots \\ \ldots\ldots\ldots\ldots\ldots\ldots\ldots\ldots \\ K(p+1),K(p+2),K(p+3),\ldots. \end{bmatrix}$$

and

$$\tilde{a}(z) = \sum_{0}^{p} \tilde{A}(i)z^{i} \quad .$$

It can be shown [18], [21] that H_{p+1} has ν linear independent rows (and no more), where ν is the degree of $\det \tilde{a}$ in every left prime MFD $(\tilde{a},\tilde{b})$. ν is called the *order* of the system. Furthermore, by the block Hankel structure of H_{p+1}, if the i-th row (column) of H_{p+1} is in the linear span of its preceding rows (columns), then so is the (i+s)-th row (column). If H_{p+1} has ν linear independent rows then seeking for the first basis rows of H_{p+1} (spanning the row space of H_{p+1}) defines integers $\nu_1,\ldots,\nu_s$ such that these first basis rows are in positions $1, 1+s,\ldots,1+(n_1-1)s, 2, 2+s,\ldots,(n_2-1)s, \ldots,s, 2s,\ldots,n_s \cdot s$, where $\nu = \nu_1 + \ldots + \nu_s$; $\alpha = (\nu_1,\ldots,\nu_s)$ are called the *dynamical indices* of the system. We take p as $\max \nu_i$. Expressing the rows of H_{p+1} in positions $1+n_1 \cdot s,\ldots,s+n_s \cdot s$ as linear combinations of their *preceding* basis rows defined above, by (2.5) gives a uniquely defined $\tilde{a}$, and thus by

$$\tilde{b} = \tilde{a} \cdot \tilde{k} \tag{2.6}$$

a unique MFD $(\tilde{a},\tilde{b})$.

Let $\tilde{a}_{ij}$ and $\tilde{b}_{ij}$ denote the (i,j) element of $\tilde{a}$ and $\tilde{b}$ respectively. For a given equivalence class, Echelon form is completely characterized by the following properties [10]:

$$\begin{aligned} &(\tilde{a},\tilde{b}) \text{ is rel. left prime; } \tilde{a}_{ii} \text{ are monic polynomials} \\ &\delta(\tilde{a}_{ij}) \le \delta(\tilde{a}_{ii}),\ j \le i;\ \ \delta(\tilde{a}_{ij}) < \delta(\tilde{a}_{ii}),\ j > i \\ &\delta(\tilde{a}_{ji}) < \delta(\tilde{a}_{ii}),\ j \ne i;\ \ \delta(\tilde{b}_{ij}) < \delta(\tilde{a}_{ii}) \end{aligned} \tag{2.7}$$

(where δ is used to denote the degree of the polynomial indicated) and conversely, every MFD satisfying (2.7) is in Echelon form. Furthermore $\nu_i = \delta(\tilde{a}_{ii})$. For a given α, the free "parameters" of a MFD in Echelon form can be described as an element θ in $\mathbb{R}^n$ where

$$n = \sum_{i=1}^{s} \delta(\tilde{a}_{ii}) + \sum_{i,j\ j<i} \{\min(\delta\tilde{a}_{jj},\delta\tilde{a}_{ii}) + \min(\delta\tilde{a}_{jj},\delta\tilde{a}_{ii}+1)\} \tag{2.8}$$

Again we identify θ with the corresponding MFD (a,b). Let $\theta_\alpha^{(2)} \subset \mathbb{R}^n$ denote the set of MFD's satisfying (2.7) for given α and let $U_\alpha^{(2)}$ denote the corresponding set of transfer functions k.

Let $M(\nu)$ denote the set of all transfer functions corresponding to systems of order ν, i.e.

$$M(\nu) = \bigcup_{\nu_1+\ldots+\nu_s=\nu} U_\alpha^{(2)} \tag{2.9}$$

Then $M(\nu)$ can be parametrized in the following way: Let $U_\alpha^{(3)}$ denote the subset of $M(\nu)$ where the rows of H_{p+1} in positions $1, 1+s,\ldots,1+(n_1-1)s, \ldots,s, 2s,\ldots,n_s\cdot s$ are basis rows of H_{p+1}, but not necessarily the first ones. Then $\{U_\alpha^{(3)} \mid \Sigma\nu_i = \nu\}$ is an overlapping covering of $M(\nu)$. Expressing the rows of H_{p+1} in position $(i+n_is)$ by the basis rows mentioned above, via (2.5) and (2.6) gives a MFD $(\tilde{a},\tilde{b})$ which is uniquely defined with respect to $U_\alpha^{(3)}$ and which has the following properties:

$$\begin{aligned} &(\tilde{a},\tilde{b}) \text{ is left prime, } \tilde{a}_{ii} \text{ are monic} \\ &\delta(\tilde{a}_{ji}) < \delta(\tilde{a}_{ii})(=\nu_i)\ , \quad j \neq i \end{aligned} \tag{2.10}$$

In these MFD's $(\tilde{a},\tilde{b})$ however not all parameters, not explicitly restricted to zero or one, are free. A vector $\theta \in \mathbb{R}^n$, $n = 2\nu s$ of free parameters for these MFD's is given by $\tilde{a}_{ij}(u)$, $u=0,1,\ldots,n_j-1$, $j,i=1,\ldots,s$; $b_{ij}(u)$, $u=0,1,\ldots,n_i-1$, $i,j=1,\ldots,s$ where $\tilde{a}_{ij}(u)$ and $\tilde{b}_{ij}(u)$ are the coefficients of $\tilde{a}_{ij}$ and $\tilde{b}_{ij}$ respectively corresponding to power z^u (see [8]). By $\theta_\alpha^{(3)}$ we denote the set of all such θ's corresponding to $k \in U_\alpha^{(3)}$. $M(\nu)$ can be shown to be a real analytic manifold and the $U_\alpha^{(3)}$, $\Sigma\nu_i = \nu$ serve as coordinate neighborhoods for this manifold [3], [16], [17].

3. TOPOLOGICAL AND GEOMETRIC PROPERTIES OF THE PARAMETRIZATIONS. For inference some topological and geometric properties of the parameter-spaces considered and of the corresponding relations between transfer functions and parameters are important.

Let θ_α denote either $\theta_\alpha^{(1)}$, $\theta_\alpha^{(2)}$ or $\theta_\alpha^{(3)}$; analogously let U_α stand for $U_\alpha^{(1)}$, $U_\alpha^{(2)}$ or $U_\alpha^{(3)}$. It turns out that most of the properties shown below do not depend on the special parameter-space considered and can be shown for more general parameter-spaces θ_α which are characterized by the two following properties [6]:

(i)   We only consider sets of MFD's with *bounded degrees*. Let $\alpha =$
      $(n_1,\ldots,n_k)$ denote the k-tuple of integer valued parameters which are
      needed to prescribe the coordinates for the vectors $\theta \in \mathbf{R}^n$ of free
      parameters for these MFD's. Thus the set of MFD's considered can be
      identified with a set $\theta_\alpha \subset \mathbf{R}^n$. We do also assume that every $\theta \in \mathbf{R}^n$
      (via the restrictions determining the "dependent" parameters) uniquely
      defines a corresponding (a,b). Furthermore, let π_α be the restriction
      of π to $\mathbf{R}^n$.

For every set of MFD's with bounded degrees the relation $b = a \cdot k$ between
transfer functions and parameters can be written as:

$$0 = (A(o),A(1),\ldots,A(p),B(1),B(2),\ldots,B(q)) \cdot \hat{K} \qquad (3.1)$$

where

$$
\hat{K} =
\left\{
\begin{array}{l}
\text{p blocks} \\[2em]
\text{q blocks}
\end{array}
\right.
\begin{pmatrix}
K(1),K(2)\ldots\ldots\ldots|K(q+1),K(q+2)\ldots\ldots\ldots K(N) \\
I\ \ ,K(1)\ldots\ldots\ldots|K(q),\ \ K(q+1)\ldots\ldots\ldots K(N-1) \\
0\qquad\qquad\qquad\ | \\
\vdots\qquad\qquad\qquad| \\
0\ldots 0..I..K(1)\ldots|K(q-p+1),K(q-p+2),\ldots..K(N-p) \\
- - - - - - - - -\,| - - - - - - - - - - - - - - - \\
-I\ \ 0\ldots\ldots\ldots..0\,| \\
0\ -I\ldots\ldots\ldots..0\,| \\
\ldots\ldots\ldots\ldots\ldots|\qquad\qquad 0 \\
0\ \ 0\ldots\ldots\ldots..-I\,|
\end{pmatrix}
$$

$$\underbrace{\qquad\qquad\qquad}_{\text{q blocks}}$$

for a suitably chosen p,q,N.

The second property required for θ_α is:

(ii) The free parameters $\theta \in \theta_\alpha$ can be obtained as unique solution from a
      subset

$$t_\alpha = \theta \cdot T_\alpha \qquad (3.2)$$

of the equations (3.1) compare e.g. [5], [8], [24]. Thereby

$$t_\alpha = (t_{\alpha_1},\ldots,t_{\alpha_s})\ ; \qquad T_\alpha =
\begin{pmatrix}
T_{\alpha_1} & 0\ldots\ldots 0 \\
0 & T_{\alpha_2}\ldots..0 \\
\ldots\ldots\ldots\ldots\ldots\ldots \\
0\ldots\ldots..0 & T_{\alpha_s}
\end{pmatrix}$$

and the rows of T_{α_i} are selected rows of $\hat{K}$ and t_{α_i} are also rows of $\hat{K}$. The selection of these rows is given by the prescription of the position of the free parameters by α. A property of such a selection is that whenever $t_\alpha \in spT_\alpha$ (i.e. t_α is a linear combination of the rows of T_α), then k is uniquely determined from t_α, T_α. Now the main point in the second part of the characterization of θ_α is that $U_\alpha = \pi_\alpha(\theta_\alpha)$ is the set *of all transfer functions* k *where* $t_\alpha \in spT_\alpha$ *and where the rows of* T_α *are linearly independent.* This property is easily checked for the three special cases discussed above, as in every case the parameter-space considered is the largest subset of the corresponding Euclidean space which is identifiable (see also [5]).

As easily seen, also the other properties are fulfilled for $\theta_\alpha^{(i)} = 1,2,3$. Moreover a much larger class of parameter-spaces originating from structural identifiability, or from canonical forms or from overlapping descriptions of $M(\nu)$ do also satisfy the requirements (i) and (ii).

If $\alpha = (n_1,\dots,n_k)$ and $\alpha^* = (n_1^*,\dots,n_k^*)$ then $\alpha \geq \alpha^*$ is used to denote $n_i \geq n_i^*$, $i = 1,\dots,k$ and $\alpha > \alpha^*$ is used if at least one strict inequality holds. If A is a set in a topological space, its closure is denoted by $\overline{A}$. Subsets of $\mathbb{R}^n$ are endowed with the corresponding natural topology. Sets of transfer functions (like U_α) will be endowed with the relative topology in the product space $(\mathbb{R}^{s\times s})^{\mathbb{N}}$ for their power series coefficients $(K(i))$, $i \in \mathbb{N}$. This topology is called the "pointwise topology" T_{pt}.

For statements similar to the two theorems below see [3], [5], [6], [7], [8], [15], [16], [17].

THEOREM 1.

(i) Θ_α *is open and dense in* $\mathbb{R}^n$

(ii) $\pi_\alpha(\overline{\Theta}_\alpha) = \bigcup_{\beta\leq\alpha} U_\alpha$

(iii) *For every* $k \in \pi_\alpha(\mathbb{R}^n)$ *the corresponding equivalence class* $\pi_\alpha^{-1}(k) \subset \mathbb{R}^n$ *is an affine subspace of dimension* (n-*rank* T_α).

Proof:

(i) These properties can be proved for $\Theta_\alpha^{(1)}$, $\Theta_\alpha^{(2)}$, and $\Theta_\alpha^{(3)}$ as in [5], [7].

(ii) If $\theta \in \mathbb{R}^n - \Theta_\alpha$, then T_α has rank less than n and $t_\alpha \in spT_\alpha$. Thus a suitable selection of rows of T_α can be performed, such that the resulting matrix T_β, $\beta < \alpha$, together with a suitable chosen t_β, $t_\beta \in spT_\beta$ uniquely define k. Thus $\pi_\alpha(\theta) \in U_\beta$. Conversely if $k \in U_\beta$, then by adding zero components to the vector $\theta \in \Theta_\beta$, this vector can be embedded in $\overline{\Theta}_\alpha$.

(iii) is easily seen from (3.2).

The function $\psi_\alpha : U_\alpha \to \Theta_\alpha$ such that $\psi_\alpha(\pi_\alpha(\theta)) = \theta$ is called the *parametrization* of U_α.

THEOREM 2.

(i) $\psi_\alpha : U_\alpha \to \Theta_\alpha$ *is a* (T_{pt} -) *homeomorphism*

(ii) U_α *is open in* $\overline{U}_\alpha$

(iii) $\pi_\alpha(\overline{\Theta}_\alpha) \subset \overline{U}_\alpha$ *and equality holds for* $s = 1$.

Proof:

(i) Clearly ψ_α is bijective; the continuity of π_α is easily seen. Thus it remains to show the continuity of ψ_α: If k, $k_t \in U_\alpha$ and $k_t \to k$ (in T_{pt}) and, therefore, using an obvious notation $(T_{\alpha,t}, t_{\alpha,t})$ (T_α, t_α), then we have from (3.2) $\theta_t = \psi_\alpha(k_t) \to \theta = \psi_\alpha(k)$, since there is a nonsingular $n \times n$ submatrix of T_α and the corresponding submatrices of $t_{\alpha,t}$ are nonsingular too from a certain t_0 onwards.

(ii) Let $k \in U_\alpha$. Then there is a neighborhood of k where T_α still has rank n. As for every $k \in U_\alpha$, $\begin{pmatrix} t_\alpha \\ T_\alpha \end{pmatrix}$ has rank less than $n+1$, the same is true for every $k \in \overline{U}_\alpha$. Thus there is a neighborhood in $\overline{U}_\alpha$ for every $k \in U_\alpha$ where T_α has rank equal to n, whereas $\begin{pmatrix} t_\alpha \\ T_\alpha \end{pmatrix}$ has rank less than $n+1$; thus this neighborhood is contained in U_α and U_α is open.

(iii) The first part is clear from the continuity of π_α. The statement for $s = 1$ follows from the structure of $\begin{pmatrix} t_\alpha \\ T_\alpha \end{pmatrix}$ (see e.g. [5]).

4. ESTIMATION. In this section we want to show how the results of the previous section are related to estimation.

For given α, estimation of θ and Σ is usually performed by the method of maximum-likelihood estimation or some of its approximations. Although we do not assume that the observations $y_T' = (y'(1),\ldots,y'(T))$ are Gaussian, the likelihood is written down as if they were Gaussian. Also (1.8) and (1.9) have to be assumed here. Then $-2T^{-1}$ by the log of the likelihood is given up to a constant by:

$$\hat{L}_T(\tau) = T^{-1} \log\det \Gamma_T(\tau) + T^{-1} y_T' \Gamma_T^{-1}(\tau) y_T \quad . \tag{4.1}$$

Thereby $\tau' = (\theta', v(\Sigma))$, where $v(\Sigma)$ denotes the vector consisting of the on- and above-diagonal elements of Σ and $\Gamma_T(\tau)$ is the $s \cdot T \times s \cdot T$ variance-covariance matrix of an ARMA process with parameters τ, i.e.

$$\Gamma_T(\tau) = \left(\int_{-\pi}^{\pi} e^{i\lambda(r-t)} k(e^{-i\lambda}) \Sigma k^*(e^{-i\lambda}) d\lambda\right)_{r,t=1,\dots,T}$$

where k is determined by θ and $*$ denotes Hermite-conjugation.

The maximum likelihood estimators then are obtained by minimizing $\hat{L}_T$. Note that $\hat{L}_T$ only depends on τ via $\Gamma_T(\tau)$ thus via k and Σ; thus (4.1) can be written as

$$\hat{L}_T(\tau) = L_{1,T}(\pi_\alpha(\theta), v(\Sigma)) \tag{4.2}$$

where $L_{1,T}$ attaches to (k,Σ) the value of the likelihood. It turns out to be convenient to extend $L_{1,T}$ to a likelihood, L_T say, defined on $\overline{U}_\alpha \times \{v(\Sigma)|\Sigma > 0\}$. L_T is a "coordinate-free" likelihood as it depends on θ only through k. For this coordinate-free likelihood the following consistency result can be shown (see [9], for similar results see [2], [14], [19], [23]): If $\lim \frac{1}{T}\sum_{t=1}^{T} y(t+s)y'(t) = Ey(t+s)y'(t)$ as stated for all $s \in \mathbb{Z}$ and if $k_0 \in \overline{U}_\alpha$ (where now also (1.8) and (1.9) hold), then the maximum likelihood estimators $\hat{k}_T$ and $\hat{\Sigma}_T$ are strongly consistent for the true k_0 and Σ_0.

If a sequence $k_T \in \overline{U}_\alpha$ converges to $k_0 \in \overline{U}_\alpha$ (in T_{pt}) then three different cases for the corresponding parameter-estimates may be distinguished:

(i) If $k_0 \in U_\alpha$, then by theorem 2 (ii), k_T will be in U_α from a certain t_0 onwards. Then also $\theta_T = \psi_\alpha(k_T)$ is defined for $T \geq t_0$ and by the continuity of ψ_α (theorem 2(i)) $\psi_\alpha(k_T)$ will converge to $\psi_\alpha(k_0) = \theta_0$ and thus we have strong consistency for the parameter-estimators.

In this case the "ideal likelihood function" (i.e. when the empirical second moments in the frequency domain form of the likelihood (4.1) have been replaced by the true ones; for more details see [9]):

$$L(\tau) = \log\det \Sigma + (2\pi)^{-1} \cdot \int_{-\pi}^{\pi} \operatorname{tr}(f^{-1}(\lambda;\tau) f(\lambda;\tau_0)) d\lambda \tag{4.3}$$

(where $f(\lambda;\tau)$ is the spectral density of an ARMA process with parameters τ and τ_0 is the true parameter) has a unique minimum at $\tau = \tau_0$. ($\hat{L}_T$ converges for $T \to \infty$ to L; for a more precise statement and the domain of convergence see [9]).

(ii) If $k_0 \in \pi_\alpha(\overline{\Theta}_\alpha) - U_\alpha$ then this transfer function is represented by the equivalence class $\pi_\alpha^{-1}(k_0) \in \mathbb{R}^n$ which by Theorem 1 is an affine subspace of dimension greater than zero. The likelihood $\hat{L}_T$ (over $\mathbb{R}^n \times \{v(\Sigma)|\Sigma > 0\}$) is constant over this equivalence class. Even if $k_T \in U_\alpha$, $T \in \mathbb{N}$, such that the corresponding parameter-estimates $\psi_\alpha(k_T) = \theta_T$ are (uniquely) well defined, not very much can be said about the sequence θ_T, as it may not converge to the true equivalence class (in the usual sense) [5]. However, if suitable prior bounds (intersecting the true equivalence class) are imposed on the norm of the elements of $\overline{\Theta}_\alpha$ and we are minimizing over this restricted space (and thus the elements of $\overline{U}_\alpha - \pi_\alpha(\overline{\Theta}_\alpha)$ are ruled out too), then a sequence θ_T, such that $\theta_T \in \pi_\alpha^{-1}(k_T)$, will converge to the true equivalence class: Since then, (in an obvious notation) $a_T^{-1} b_T \to k_0$ (by the boundedness of the a_T's) implies $b_T - a_T \cdot k_0 \to 0$.

For $k_0 \in \pi_\alpha(\overline{\Theta}_\alpha) - U_\alpha$ the ideal likelihood (4.3) assumes its minimum along the k_0-equivalence class. Therefore, also for calculating the minimum of the actual likelihood the algorithm will tend to be rather unstable along the corresponding directions.

(iii) Finally we consider the case $k_0 \in \overline{U}_\alpha - \pi_\alpha(\overline{\Theta}_\alpha)$. An example for this case has been given in [7]. Here, even if $k_T \in U$, the corresponding sequence θ_T must be such that $\|\theta_T\| \to \infty$ since otherwise, by a compactness argument, a subsequence converging to a point in $\mathbb{R}^n$ would exist, in contradiction to $k_T \to k_0 \in \pi_\alpha(\mathbb{R}^n)$. Thus in this case, in doing actual calculations, the computer will overflow for T sufficiently large.

Comparing the descriptions of $M(\nu)$ as

$$M(\nu) = \bigcup_{\Sigma\nu_i=\nu} U_\alpha^{(3)} \quad ; \quad M(\nu) = \bigcup_{\Sigma\nu_i=\nu} U_\alpha^{(2)}$$

note that the $U_\alpha^{(3)}$ are dense in $M(\nu)$ (see e.g. [8]) whereas the $U_\alpha^{(2)}$ in general have different dimensions and only one of them is open and dense in $M(\nu)$. Thus, if only ν is known, but not α, for every choice of α, $k_0 \in M(\nu)$ can at least be arbitrarily closely approximated by $k_t \in U_\alpha^{(3)}$. On the other hand if α is known and if $U_\alpha^{(2)}$ has lower dimension than $U_\alpha^{(3)}$, then estimating θ in $\Theta_\alpha^{(2)}$ will cause an "efficiency gain" compared with $\Theta_\alpha^{(3)}$.

In the case of the prescription of the column degrees, α is not uniquely determined from k, in the sense that $k \in U_\alpha^{(1)}$, $k \in U_\beta^{(1)}$; $\alpha \neq \beta$ is possible [5]. Also $U_\alpha^{(3)} \cap U_\beta^{(3)} \neq 0$ if the corresponding orders are same. To the contrary $\{U_\alpha^{(2)} | \alpha \in \mathbb{Z}_+^s\}$ is a partitioning of U_A and therefore in this case α is uniquely defined from k.

If α is not known a-priori, it has to be estimated too. One method of estimating α among a finite number of alternatives is to penalize the maximum likelihood estimator $\hat{\Sigma}_\alpha$ of Σ_0 (over $\overline{U}_\alpha$) with a term depending on the dimension $n(\alpha)$ of the vector θ of free parameters in the corresponding Θ_α (see e.g. [1] [13] [22]) and thus to minimize

$$A(\alpha) = \log\det \hat{\Sigma}_\alpha + \frac{C(T)}{T} \cdot n(\alpha) \tag{4.4}$$

over $\alpha \in I$, $I \subset (\mathbb{Z}_+)^k$, I finite, where $C(T)$ is prescribed a-priori. One possible choice for $C(T)$ is $\log T$. Then (4.4) is called the BIC-criterion. An information theoretic derivation of this criterion is given in [22]. For consistency properties of this and other choices of $C(T)$ (under some additional assumptions) in [13]. In slightly more general terms consistency is understood as follows: If

$$k_0 \in U_0 = \bigcup_{\alpha \in I} U_\alpha$$

then the estimators $\hat{\alpha}_T$ converge to the set J of all $\alpha \in I$ such that $k_0 \in \overline{U}_\alpha$ and $n(\alpha)$ is minimal among all α's such that $k_0 \in \overline{U}_\alpha$. For suitable C(T) this consistency is achieved (under additional assumptions) by the fact that whenever $k_0 \in \overline{U}_\alpha$, $k_0 \notin \overline{U}_\beta$ then the difference between the first terms on the right hand side of (4.4) will asymptotically dominate the difference of the second terms and if $k_0 \in \overline{U}_\alpha$, $k_0 \in \overline{U}_\beta$ then the differences in the second terms will be more important.

If $k_0 \in U_\alpha^{(3)}$, by (4.4) only the corresponding order can be estimated consistently (this has been shown in [13]) whereas for the estimation of α different approaches have to be used as $\overline{U}_\beta^{(3)} = \overline{U}_\alpha^{(3)}$ whenever the orders corresponding to α and β are the same.

REFERENCES

1. H. Akaike. *Canonical correlation analysis of time series and the use of an information criterion.* In System Identification: Advances and Case Studies (R. K. Mehra and D. G. Lainiotis, eds.), Academic Press, New York, 1976, pp. 27-96.

2. P. E. Caines. *Stationary linear and nonlinear identification and predictor set completeness.* IEEE Trans. AC 23, (1978), pp. 583-594.

3. J. M. C. Clark. *The consistent selections of parametrizations in systmes identification.* Paper presented at JACC 1976.

4. M. Deistler. *The structural identifiability of linear models with autocorrelated errors in the case of cross-equation restrictions.* J. of Econometrics 8, (1978), pp. 23-31.

5. M. Deistler. *The properties of the parametrization of ARMAX systems and their relevance for structural estimation and dynamic specification.* Paper presented at the 4th Econometric Society World Congress (1980).

6. M. Deistler. *A unified approach to some properties of the structure of ARMA systems.* Mimeo (1982).

7. M. Deistler, W. Dunsmuir, and E. J. Hannan. *Vector linear times series models: Corrections and extensions.* Adv. Appl. Prob. 10, (1978), pp. 360-372.

8. M. Deistler and E. J. Hannan. *Some properties of the parametrization of ARMA systems with unknown order.* J. of Multivariate Analysis 11 (1981), pp. 474-484.

9. W. Dunsmuir and E. J. Hannan. *Vector linear time series models.* Adv. Appl. Prob. 8 (1976), pp. 339-364.

10. D. G. Forney, Jr. *Minimal basis of rational vector spaces with applications to multivariable linear systems.* SIAM J. of Control 13 (1975), pp. 493-520.

11. E. J. Hannan. *The identification of vector mixed autoregressive moving average systems,* Biometrika 57 (1969), pp. 223-225.

12. E. J. Hannan. *The identification problem for multiple equation systems with moving average errors.* Econometrica 39 (1971), pp. 751-767.

13. E. J. Hannan. *Estimating the dimensions of a linear system.* J. of Multivariate Analysis 11 (1981), pp. 459-473.

14. E. J. Hannan, W. Dunsmuir, and M. Deistler. *Estimation of vector ARMAX models.* J. of Multivariate Analysis 10 (1980), pp. 275-295.

15. M. Hazewinkel. *Moduli and canonical forms for linear dynamical systems II: The topological case.* Math Systems Theory 10 (1977), pp. 363-385.

16. M. Hazewinkel and R. E. Kalman. *On invariants, canonical forms and moduli for linear constant finite-dimensional, dynamical systems.* Lecture Notes in Economics, Math. System Theory 131, Springer Berlin, 1976.

17. R. E. Kalman. *Algebraic geometric description of the class of linear systems of constant dimension.* 8th Ann. Princeton Conf. on Information Sciences and Systems, Princeton, N.J., 1974.

18. R. E. Kalman, P. L. Falb, and M. A. Arbib. *Topics in Mathematical System Theory.* McGraw-Hill, New York, 1969.

19. L. Ljung. *Convergence analysis of parametric identification methods.* IEEE Trans. AC 23 (1978), pp. 770-783.

20. V. M. Popov. *Invariant description of linear, time-invariant controllable systems.* SIAM J. of Control 10 (1972), pp. 252-264.

21. J. Rissanen. *Basis of invariants and canonical forms for linear dynamic systems.* Automatica 10 (1974), pp. 175-182.

22. J. Rissanen. *Modelling by shortest data description.* Automatica 14 (1978), pp. 465-471.

23. J. Rissanen and P. E. Caines. *The strong consistency of maximum likelihood estimators for ARMA processes.* Ann. Statistics 7 (1979), pp. 297-315.

24. V. Wertz and M. Gevers. *A note on different ARMA representations.* Mimeo 1981.

UNIVERSITY OF TECHNOLOGY
VIENNA
AUSTRIA

SOME GEOMETRIC METHODS FOR STOCHASTIC INTEGRATION IN MANIFOLDS*

T. E. Duncan

1. INTRODUCTION. A comparison is made of the various definitions of the real-valued integrals of processes that are formed by the pairing of 1-forms and a semimartingale in a Riemannian manifold. The comparison emphasizes a global differential geometric approach. In addition to this comparison another approach to these real-valued semimartingales is introduced which can be naturally identified with the methods that are used in Euclidean spaces and with the study of geodesics in manifolds. A basic ingredient of this approach is the use of the canonical 1-form of the linear connection.

Itô [10] defined the integral of a suitably measurable stochastic process integrand with respect to Brownian motion which preserved some of the important probabilistic properties of a Wiener integral. The basic property is that the integral is a martingale with respect to a natural increasing family of σ-algebras. Using some results from the theory of martingales, especially the Doob-Meyer decomposition, the definition of Itô can be extended to integrals with respect to other martingales, both continuous and discontinuous (e.g. [14]). The change of variables rule for Itô integrals reflects the unbounded variation property of Brownian motion so that the usual rules of calculus are not satisfied. Another definition of integrals with respect to Brownian motion has been given independently by Fisk [7] and Stratonovich [16] that satisfies the usual rules for change of variables. These latter integrals are defined for a smaller class of integrands than the Itô integrals and the relation between these two definitions is well known [16].

In the study of semimartingales in manifolds the first question that should be addressed is the existence of such processes in a manifold. On the one hand some abstract theorems for the existence of probability measures in topological spaces could be used for the construction in certain manifolds. However this approach is fairly technical, it neglects the geometry that is available and it seems to be difficult to get the best results. Another approach is to take Brownian motion or a related semimartingale in a Euclidean space and identify it with a stochastic process in a manifold by the use of geometric techniques.

---

*Research supported by NSF Grant ECS-8024917

This method has been used by Duncan [1,5] and independently Malliavin [13] has described a related approach. Methods previous to these results were local in nature and required implicitly that the manifolds satisfied special properties (e.g. [8]).

For an integrator that is a manifold-valued Brownian motion real-valued integrals have been defined [1,2,5] that are martingales with respect to a natural increasing family of σ-algebras. Similarly real-valued integrals can be defined that are the analogue of the integrals of Fisk and Stratonovich (e.g. [13]). In addition, if the manifold is a vector bundle, then discontinuous semimartingales can be introduced and real-valued integrals can be defined for these processes [4,6].

Recently Meyer [15] has introduced another approach to the integrals of Itô and Stratonovich in terms of local coordinates in the manifold using the notion of second order differentials.

In this paper the geometric relation among these various approaches to real-valued integrals formed from manifold-valued semimartingales is given. The semimartingales are restricted to Brownian semimartingales because these interact naturally with the geometry of a Riemannian manifold. A manifold-valued Brownian semimartingale is a manifold-valued process that can be described in local coordinates as a semimartingale whose martingale part is a continuous martingale whose associated increasing process is absolutely continuous with respect to Lebesgue measure and whose process of bounded variation is also absolutely continuous with respect to Lebesgue measure. By the change of variables formula of Itô it is clear that this definition does not depend on the local coordinates that are used. The subsequent results have suitable generalizations to arbitrary continuous, local, manifold-valued semimartingales.

2. GEOMETRIC FORMULATION. To relate some of the approaches to real-valued integrals of manifold-valued semimartingales it is useful to introduce the canonical 1-form θ of a linear connection. Let $L(M)$ be the bundle of linear frames of the manifold M. The canonical form θ of $L(M)$ [12] is the $\mathbb{R}^n$-valued 1-form on $L(M)$ defined by

$$\theta(X) = u^{-1}\,(\pi(X)) \tag{1}$$

for $X \in T_u(L(M))$ where u is considered as an element of $HOM(\mathbb{R}^n, T_{\pi(u)}M)$. Since the manifolds that are considered here are Riemannian, the bundle of linear frames can be reduced to the bundle of orthonormal frames, $O(M)$.

Perhaps somewhat more concretely the canonical form can be defined by the vector of 1-forms $\theta^1,\dots,\theta^n$ where

$$d\pi(X) = \sum \theta^i(\pi(X))f_i \tag{2}$$

and $u = (x, f_1, \ldots, f_n) \in L(M)$. Thus

$$\theta(X) = u^{-1}(d\pi(X)) \quad . \tag{3}$$

Let $\zeta \in \mathbb{R}^n$. For each $u \in L(M)$ let $(B(\zeta))_u$ be the unqiue horizontal vector at u such that $\pi((B(\zeta))_u) = u(\zeta)$. The vector field $B(\zeta)$ is called the standard horizontal vector field corresponding to ζ. Clearly these vector fields depend on the choice of the linear connection. By the construction of $B(\zeta)$ it is clear that $\theta(B(\zeta)) = \zeta$ and $\omega(B(\zeta)) = 0$ where ω is the connection form.

Since the vector field $B(\zeta)$ is constructed to be horizontal it is not surprising that it is directly related to geodesics. In fact the projection of any integral curve of a standard horizontal vector field of $L(M)$ is a geodesic and conversely every geodesic can be obtained in this way. For example let b_t be an integral curve of $B(\zeta)$ and let $x_t = \pi(b_t)$. Then $\dot{x} = \pi(\dot{b}_t) = \pi(B_{b_t}) = b_t\zeta$ where $b_t\zeta$ is the image of ζ by $b_t \in \mathrm{Hom}(\mathbb{R}^n, T_{x_t}M)$. The vectors of $b_t\zeta$ are parallel along x_t because b_t is a horizontal lift of x_t. Notice that in this bundle a geodesic is described by a "first-order" property, that is a property on vector fields whereas in local coordinates a geodesic is characterized by a "second-order" property, that is a second order differential equation. It is this important feature of the description of geodesics that carries over to the description of Brownian semimartingales. Furthermore it is not surprising that a differential or first-order description is directly associated with parallelism and horizontal lifts of Brownian semimartingales.

For our purposes a manifold-valued martingale is understood as an M-valued process with continuous sample paths such that the local real-valued Itô integrals formed by the pairing of this process with the family of smooth 1-forms are real-valued continuous local martingales with respect to the increasing family of σ-algebras generated by the M-valued process.

The manifolds that are considered here are always Riemannian manifolds, that is, there is a metric and the connection is the torsion free metric connection. The integral of Itô is primarily emphasized though it is shown how the Stratonovoch integral is related to the Itô integral. For Brownian motion and martingales formed from it the geometric notion of parallelism is an important technique. For example, the martingale property is preserved under parallelism. Parallelism can be viewed in the frame bundle by the horizontal lift of a manifold-valued curve. This property suggested the use of the canonical form θ of the linear connection.

PROPOSITION 1. *Let* M *be a Riemannian manifold. The canonical 1-form* θ *of the connection can be extended to be defined on the formal vectors of a Brownian semimartingale.*

Proof. For a piecewise smooth curve in the manifold the horizontal lift of it can be formed and the canonical form θ acts on this curve in the frame bundle. Since the action of $GL(n)$ (resp. $O(n)$) on the initial fiber of $L(M)$ (resp. $O(M)$) for the curve commutes with the construction of the horizontal lift θ is intrinsically defined for the curve.

Given an M-valued Brownian semimartingale there is a horizontal lift of this process to the bundle of orthonormal frames. To accomplish this lift it is only necessary to follow the same techniques as were used in [3]. Since the horizontal lift of a Brownian semimartingale is defined, θ is intrinsically defined on it as it is for piecewise smooth curves.

A similar result can be proved for an arbitrary torsion free linear connection by naturally extending the methods in [3].

The description of θ on the formal vectors of a Brownian semimartingale provides an intrinsic description of these formal vectors. With these intrinsic formal vectors, real-valued integrals can be constructed in a natural way when these formal vectors are paired with a smooth 1-form or a smooth vector field where the latter is identified canonically with a 1-form by the metric. The construction of such a real-valued integral can be related intrinsically to the computation of a similar integral in a Euclidean space. Since real-valued integrals of a semimartingale depend continuously on the initial value it suffices to consider a semimartingale that starts at a fixed point and this point can even be considered to be in the local chart where the real-valued integral is to be constructed. Since parallelism is defined for the Brownian semimartingale because the horizontal lift is defined the formal vectors of the Brownian semimartingale and the smooth 1-form or vector field along this process can be parallel transported to the fiber of the tangent bundle at the initial point. The parallel transport of the manifold-valued Brownian semimartingale to the Euclidean space defines a Brownian semimartingale in this latter space. Clearly the necessary measurability properties are preserved. In this Euclidean space the usual techniques for integration of semimartingales can be used.

A (Brownian) martingale can be defined in terms of the canonical 1-form θ. The manifold-valued process (X_t) is a martingale if the formal vectors $\theta^i(dX_t)$ are the formal vectors of a martingale. This property can be verified by the parallel translation of these vectors to the initial tangent space of the manifold-valued process and the integration of these formal vectors in this Euclidean space. Furthermore it will be shown that there is also an intrinsic local coordinate description of this martingale property. However it follows immediately from the previous description of the martingale property that all real-valued integrals formed from the pairing of smooth 1-forms or vector fields with the process are real-valued local martingales.

The parallelism definition of the martingale property can be rephased to describe a manifold-valued martingale as the integral of a family of formal vectors that are the parallel transport along the process of the formal vectors of an $\mathbb{R}^n$-valued martingale.

The use of the canonical form θ provides a "first-order" description of a Brownian martingale while the local coordinates provide a "second-order" description. This is analogous to the aforementioned description of geodesics.

If a manifold-valued Brownian motion is constructed by the extension of the inverse of the development as in [1,5] then it is clear that the real-valued integrals that are constructed with the canonical form θ from the Brownian motion are martingales.

A local coordinate description of a Brownian semimartingale can also be given instead of the global description using the canonical form θ [15]. The following is a local coordinate description of a Brownian martingale.

PROPOSITION 2. *The process* (X_t) *is an* M-*valued continuous Brownian local martingale if and only if for each patch and for some local coordinate system*

$$dx_t^i + 1/2 \sum_{j,k} \Gamma_{jk}^i \, d\langle X^j, X^k\rangle_t = dY_t^i \tag{4}$$

where dY_t^i $i = 1,2,\ldots,n$ *are the differentials of* n *real-valued Brownian martingales and* Γ_{jk}^i *are the components of Christoffel symbols of the linear connection.*

Proof. ($\Rightarrow$) Let $x^1,\ldots,x^n$ be a local coordinate system. Consider a local trivialization of the bundle of linear frames given by these local coordinates. Compute θ from this local trivialization. Since (X_t) is a Brownian martingale $\theta^i(\pi(d\tilde{X}_t))$ is the differential of a real-valued martingale where $(\tilde{X}_t)$ is the (local) horizontal lift of (X_t). The equation (4) is a collection of differentials of n real-valued Brownian martingales because $\theta^i(\pi(d\tilde{X}_t))$ is the differential of a real-valued Brownian martingale. The description (4) is invariant because another local coordinate system would merely describe another trivialization of the bundle of linear frames.

($\Leftarrow$) Conversely let $(x^1,\ldots,x^n)$ be the local coordinate system such that $dY_t^1,\ldots,dY_t^n$ are the differentials of n real-valued continuous local Brownian martingales. Consider the trivialization of the bundle of linear frames that is given by this local coordinate system. The differential description (4) is an invariant description by the change of variables formula as it was applied in the other half of the proof. Since these differentials provide an invariant description they can be "lifted" to the frame bundle. Furthermore $\theta^i(\pi(d\tilde{X}_t))$ is the differential of a Brownian martingale.

Another definition of integration of Brownian semimartingales is useful because it obeys the usual laws of coordinate change for smooth curves. This is the definition that was introduced in Euclidean spaces by Fisk [7] and Stratonovich [16]. Since it should obey the usual rules of calculus it should only depend on the notion of the tangent to a curve which can be made without the notion of a connection. However it should be recalled to construct a Brownian semimartingale it is necessary to have a Riemannian connection.

Let $\alpha = \sum a_i dx^i$ be a smooth 1-form in M. The symmetric or Fisk-Stratonovich integral of α with respect to a (Brownian) semimartingale is defined in local coordinates as

$$\sum_{i,j} \int a_i(X_s)dX_s^i + 1/2 \int D_i a_j d\langle X^i, X^j\rangle$$

where the first integral is an Itô integral and D_i is differentiation in the direction x^i.

Meyer [15] has shown the intrinsic property of the Stratonovich integral by introducing the notion of a second order differential form. Our justification of the intrinsic property of the Stratonovich integral uses the same techniques as Meyer with a slightly different formulation. Define an action $\tilde{d}$ on 1-forms as

$$\tilde{d}\alpha = \sum a_i dx^i + \sum_{i,j} D_j a^i dx^j dx^i$$

and a product

$$\alpha \cdot \beta = \sum_{i,j} a_i b_j dx^i dx^j$$

where $\alpha = \sum \alpha_i dx^i$ and $\beta = \sum b_i dx^i$. The product is commutative and bilinear over the ring $C^\infty(M)$. Furthermore

$$\tilde{d}(f\alpha) = f\tilde{d}\alpha + df \cdot \alpha$$

where $f \in C^\infty(M)$. These properties show that the definition of $\tilde{d}$ is intrinsic.

The Stratonovich integral has a nice functorial property because $\tilde{d}$ possesses a nice functorial property, that is, it maps naturally under pullbacks. Let $F : M \to N$ be a smooth map between manifolds M and N and let α be a 1-form on N. Let $Z_t = F(X_t)$ where (X_t) is an M-valued process. Then

$$\int_{Z_0^t} \alpha = \int_{X_0^t} F^*(\alpha)$$

where these are Stratonovich integrals. This fact follows from $F^*(\tilde{d}\alpha) = \tilde{d}F^*(\alpha)$.

Similarly it is natural to examine the functoriality properties of the Itô integral. Consider a real-valued martingale formed from Brownian motion and a smooth 1-form. Since the martingale property is intimately related to the connection it is reasonable to expect that it is difficult to preserve this property under a mapping. Of course this fact is well known in Euclidean spaces. For the map $F: M \to N$ one can directly compute terms in local coordinates to obtain the condition for the preservation of the martingale property [15]. However a geometric explanation will be provided which is in the spirit of the geometric approach to the Itô integrals using the canonical 1-form of the connection.

Initially consider a standard horizontal vector field. Recall that the integral curves of such vector fields are all the geodesics. Consider the derivative of $F\,\nabla F_x : T_xM \to T_{f(x)}N$. The derivative ∇F is a section of the bundle $L(TM,F^*TN)$ where F^*TN is the pull-back of TN to a bundle over M. The second derivative $\nabla\nabla F$ is the derivative of ∇F with respect to the natural connection on $L(TM,F^*TN)$. Thus $\nabla\nabla F$ is a section of the bundle $L_S^2(TM,F^*TN)$ of symmetric bilinear maps. The connection on $L(TM,F^*TN)$ is given as the negative of the connection on TM plus the connection on F^*TN. This rule for combination follows from the usual pairing methods. The pull-back connection of F^*TN is given in local coordinates as

$$\Gamma^{\alpha}_{\beta\gamma}(F(x))\,\frac{\partial F^{\beta}}{\partial x^{i}}$$

where $\Gamma^{\alpha}_{\beta\gamma}$ are the connection coefficients or Christoffel symbols of the connection on N. This pull-back connection is the obvious quantity given ∇F.

Consider the condition to preserve a standard vector field $B(\zeta)$. Since $B(\zeta)$ can be described in terms of the covariant derivative, the condition for the preservation of the horizontal property is $\nabla_X\nabla_X F = 0$. Now consider the formal vectors of a Brownian motion. Recall that the probability law for Brownian motion is invariant under the orthogonal group. Using this orthogonal group invariance with the above condition for standard horizontal vector fields the condition that preserves the martingale property is

$$\mathrm{Tr}\nabla\nabla F = 0 \quad .$$

By definition this is the Laplacian ΔF.

As Meyer [15] has noted the intrinisic geometric description of the Itô and the Stratonovich integrals provides an intrinsic description of the difference between these two integrals. Let I_I and I_S be the Itô and the Stratonovich integrals respectively with respect to a Brownian semimartingale. In local coordinates the difference between these two integrals is

$$I_I(\alpha) - I_S(\alpha) = 1/2 \sum \int [a_k(X_s)\Gamma^k_{ij}(X_s) - D_i a_j(X_s)]\, d\langle X^i, X^j\rangle_s \tag{4}$$

where $\alpha = a_i dx^i$. If the integrator is Brownian motion then the difference is

$$I_I(\alpha) - I_S(\alpha) = 1/2 \int \delta\alpha(X_s)\, ds \tag{5}$$

where δ is the adjoint of d, the exterior derivative. This result (5) for the difference between the Itô and the Stratonovich integrals can be obtained from the known result in Euclidean spaces and some elementary geometry properties. To explain briefly this approach recall that the derivative of a function with respect to a vector field is the action of the vector field on the function, that is,

$$\nabla_X f = Xf$$

where $f \in C^\infty(M)$.

The development and its inverse extended to the paths of Brownian motion provide an almost sure isomorphism between an M-valued Brownian motion starting at $a \in M$ and a T_aM-valued Brownian motion [1,5]. Furthermore the real-valued Itô integrals formed from the Brownian motion can be defined using this parallelism [1,2].

Similarly the Stratonovich integral can be defined by this parallelism method. This can be accomplished by following the proof in [1,5] using the Stratonovich definition of integral instead of the Itô definition. The smoothness of the integrand suffices to verify this method.

To show that the difference between these two integrals computed in T_aM is (5) let $\alpha = \sum a_i dx^i$ be a smooth 1-form.

In T_aM these integrals can be represented by the following pairing

$$\int \langle \tau_0^t \alpha, \tau_0^t \, dB_t \rangle$$

where τ_0^t is the parallelism along the Brownian paths such that $\tau_0^t : T_{B_t}M \to T_aM$ and α is considered as a vector field using the canonical identification from the metric. The above integral can be written as

$$\int \langle \tau_0^t \alpha, d\tilde{B}_t \rangle$$

where $(\tilde{B}_t)$ is a T_aM-valued Brownian motion.

Now

$$\int \langle \alpha, dB_t \rangle = \int \langle \alpha, (\tau_0^t)^t d\tilde{B}_t \rangle$$

defines the formal vectors dB_t of a Brownian motion because the parallelism is an action of the orthogonal group that has the appropriate measurability property. The Itô and the Stratonovich integrals satisfy a well known relation in Euclidean spaces [16]. The terms in this expression are the derivatives of functions in the coordinate directions in T_aM which are the action of these coordinate vector fields on the functions. However these computations are the same as in $T_{B_t}M$ as was noted above. Thus the relation between the Itô and the Stratonovich integrals in manifolds follows immediately from the known results in Euclidean spaces. This result can be described more picturesquely by considering an observer along the paths of the Brownian motion. An observer relating these two definitions of integrals in an orthonormal frame along the Brownian motion performs the same computation as in Euclidean space and he can detect no difference from the computation.

3. APPLICATIONS. The canonical 1-form θ for the connection can be used to describe stochastic differential equations in a useful intrinsic manner. This formalism was used in [1,2,3] without the explicit use of θ. However it seems worthwhile to review these ideas using θ because of its usefulness in geometry. This description is easily accomplished if a local trivialization of the bundle of orthonormal frames is obtained in a convex neighborhood by parallel transporting an orthonormal frame along the unique local geodesics. A stochastic differential equation can be represented intrinsically as

$$\theta(dX_t) = adt + bdB_t \tag{6}$$

for a diffusion type equation. Using the previous techniques this equation can be represented in local coordinates.

Transformation of measures by absolute continuity methods can be used to define manifold-valued Brownian semimartingales [1,2,5].

PROPOSITION 3. *Let* (X_t) *be an* M-*valued Brownian motion on the probability space* (Ω,F,P) *where* M *is a compact Riemannian manifold. Let* (N_t) *be*

$$N_t = \exp\left[\int_0^t \langle\alpha_s, dX_s\rangle - 1/2\int_0^t \langle\alpha_s,\alpha_s\rangle ds\right]$$

where α *is a smooth vector field and the first integral in the exponential is an Itô integral. Then* $d\tilde{P} = NdP$ *is a probability measure and*

$$\theta(dX_t) = \alpha_t dt + dB_t$$

where (B_t) *is a* $(\Omega,F,\tilde{P})$ *Brownian motion.*

Such a result has also been described by the Stratonovich integral [9].

REFERENCES

1. T. E. Duncan, *Some stochastic systems on manifolds*, Lecture Notes in Econ. and Math. Systems, 107 (1975), 262-270, Springer-Verlag.

2. T. E. Duncan, *Dynamic programming optimality criteria for stochastic systems in Riemannian manifolds*, Appl. Math. and Optimization, 3 (1977), 191-208.

3. T. E. Duncan, *Some filtering results in Riemannian manifolds*, Information and Control, 35 (1977), 182-195.

4. T. E. Duncan, *Estimation for jump processes in the tangent bundle of a Riemannian manifold*, Appl. Math. and Optimization, 4 (1978), 265-274.

5. T. E. Duncan, *Stochastic systems in Riemannian manifolds*, J. Optimization Th. and Appl., 27 (1979), 399-426.

6. T. E. Duncan, *Optimal control of continuous and discontinuous processes in a Riemannian tangent bundle*, Lecture Notes in Math., 794 (1980), 396-411, Springer-Verlag.

7. D. L. Fisk, *Quasi-martingales*, Trans. Amer. Math. Soc. 120 (1965), 369-389.

8. R. Gangolli, *On the construction of certain diffusions on a differentiable manifold*, Z. Wahrscheinlichkeitstheorie 2 (1964), 406-419.

9. N. Ideka and S. Manabe, *Integral of differential forms along the path of diffusion processes*, Publ. RIMS, Kyoto Univ., 15 (1979), 827-852.

10. K. Itô, *Stochastic integral*, Proc. Imperial Acad., Tokyo 20 (1944), 519-524.

11. K. Itô, *On a formula concerning stochastic differentials*, Nagoya Math. J. 3 (1951), 55-65.

12. S. Kobayashi and K. Nomizu, *Foundations of Differential Geometry*, Vol. I, Interscience 1963.

13. P. Malliavin, *Géométrie Differentielle Stochastique*, Presses de l'Université de Montréal, 1978.

14. P. A. Meyer, *Intégrales stochastiques I. Séminaire de probabilités 1.* Lecture Notes in Math., 39 (1967), 72-117, Springer-Verlag.

15. P. A. Meyer, *A differential geometric formalism for the Itô calculus*, Lecture Notes in Math. 851 (1981), 256-270, Springer-Verlag.

16. R. L. Stratonovich, *A new representation for stochastic integrals and equations*, Vestnik Moskov. Univ. Ser. I. Mat. Meh., 1 (1964), 3-12 (translated in SIAM J. Control 4 (1966), 362-371).

DEPARTMENT OF MATHEMATICS
UNIVERSITY OF KANSAS
LAWRENCE, KANSAS 66045
U.S.A.

OVERLAPPING PARAMETRIZATIONS FOR THE REPRESENTATION OF MULTIVARIATE STATIONARY TIME SERIES*

Michel Gevers and Vincent Wertz

ABSTRACT. When identifying a (state-space or ARMA) model for a multivariate stationary stochastic process using a black-box approach, the first problem consists in finding the order of the model and defining a uniquely indentifiable parametrization. For a given order, several "overlapping" parametrizations, all involving the same number of parameters, can usually be fitted to the same process. The question then arises as to whether some parametrizations are better than others. We present an asymptotic result showing that all overlapping parametrizations give the same value to the determinant of the Fisher information matrix and, therefore, with many identification schemes, to the determinant of the asymptotic error covariance matrix. For finite data, some structures may still be better than others, and two heuristic structure estimation methods are analyzed. Some simulation results are also presented.

1. INTRODUCTION. An important and widely studied problem in the theory of identification of multivariate stationary finite-dimensional stochastic processes is that of determining the structure of the state-space or ARMA model for that process such that the model parameters become uniquely identifiable. Two different lines of thought have been followed for this problem. The first idea is to use canonical (state-space or ARMA) forms [1]-[5]. To any finite dimensional process one can associate a canonical form in a unique way by specifying a selection procedure. Different selection procedures will lead to different canonical forms, but the parameters in any two canonical representations of a given process are related by a bijective relationship. The structure of a canonical representation of a process is determined by a set of "structural invariants" (e.g. the Kronecker invariants) which are again uniquely defined by the process and the selection procedure. The disadvantage with using canonical forms is that the estimation of those structural invariants is very critical: if they are wrongly estimated, then the parameter estimation problem becomes ill-conditioned.

---

*This work was performed in part while the authors were on leave at the University of Newcastle, New South Wales, Australia. The work was supported by IRSIA (Belgium) and the Australian Research Grants Committee.

In recent years an alternative approach has been proposed, namely that of using "overlapping parametrizations" [6]-[12]. It has been recognized that the set of all finite dimensional systems can be represented by a finite number of parametrizations, each parametrization being uniquely identifiable. To each parametrization there corresponds a set of integers called "structure indices." Each of these parametrizations also called "structure" is able to represent almost all finite dimensional systems, each system can normally be represented by more than one such parametrization, and any two parametrizations for a given process are related by a linear transformation which corresponds to a coordinate transformation in Euclidean space; hence the use of the word "overlapping" parametrizations. Now because a process can be represented in more than one overlapping form, the question naturally arises as to whether, for a given data set, any such form is better than the others in a statistical or numerical sense. More specifically, given an observation record $\{y_t, t = 0,1,\ldots,N\}$ generated by a finite dimensional stochastic system which can be represented by different overlapping structures, we want to know whether one particular structure will yield more efficient parameter estimates than the others, or whether using one particular structure will produce a numerically better conditioned algorithm. This question has been considered by several authors in recent years, but so far no definite answer is available. Different procedures have been proposed that select one out of several candidate parametrizations which is considered best in some ad hoc sense [7]-[9]. The "optimal" parametrization is related in a rather intuitive way to the precision with which the parameters are estimated. In [8] a different approach has been taken: the idea is not to select a best structure, but to change from one parametrization to a better one (by a coordinate transformation) when the parameter estimation algorithm runs into numerical difficulties.

In this paper we first show that, for a n-th order process with a p-dimensional observation vector y_t, the parameters of any overlapping parametrization (either in state-space or in ARMA form) are obtained from a set of 2np "intrinsic invariants" which are determined from the Hankel matrix of impulse responses. To any choice of p structure indices, there corresponds, for a given system, a set of 2np parameters which completely specify this system. The choice of the p structure indices determines in which particular local coordinate space the system is described. From these 2np "intrinsic invariants," a unique state-space or ARMA parametrization can then be derived; these will belong to the set of overlapping parametrizations.

Next we compare the different overlapping parametrizations in terms of asymptotic accuracy of the parameter estimates. We show that, if the determinant of the Fisher information matrix is used as a measure of asymptotic efficiency, then all overlapping parametrizations describing the same process are equivalent, in the sense that they will give the same value to this criterion.

Our result implies that if a process is modelled in state-space or ARMA form using a prediction error method, then the determinant of the covariance matrix of the parameter estimates will asymptotically be the same, whichever overlapping parametrization is used. This does not mean that, over a finite data record, one parametrization might not have a better numerical behavior than others. We proposed two heuristic methods to deal with this finite data problem. Some simulation results are also presented.

2. PARAMETRIZATION OF MULTIVARIATE SYSTEMS. We consider throughout this paper a p-dimensional stationary full rank zero-mean stochastic process $\{y_t\}$ with rational spectrum. Then it is well known that $\{y_t\}$ can be described, up to second order statistics, by the following finite-dimensional representations:

STATE-SPACE REPRESENTATION

$$x_{t+1} = Fx_t + Ke_t$$

$$y_t = Hx_t + e_t \tag{2.1}$$

where the state x_t is an n-dimensional vector; F, K and H are matrices of dimensions $n \times n$, $n \times p$, and $p \times n$, F has all its eigenvalues strictly inside the unit circle, and $\{e_t\}$ is a p-dimensional white noise sequence with covariance matrix Q.

INPUT-OUTPUT REPRESENTATION (ARMA Model)

$$y_t + A_1 y_{t-1} + \dots + A_r y_{t-r} = e_t + B_1 e_{t-1} + \dots + B_s e_{t-s} \tag{2.2a}$$

where $A_1, \dots, A_r$, $B_1, \dots, B_s$ are $p \times p$ matrices and $\{e_t\}$ is as before. This representation is equivalent with the following:

$$A(z)y_t = B(z)e_t \tag{2.2b}$$

where A(z) and B(z) are square polynomial matrices in the variable z (z is the advance operator: $zy_t = y_{t+1}$), with $\det A(z) \neq 0$ for $|z| \geq 1$, and $\lim_{z\to\infty} A^{-1}(z)\ (z) = I$.

Without any loss of generality, we can make the following assumptions regarding these two representations.

ASSUMPTION 1a: *The matrix triple* (H,F,K) *is of minimal order* n, *where* n *is the dimension of the state vector* x_t, *i.e.*

$$\text{rank}\begin{bmatrix} H \\ HF \\ \cdot \\ \cdot \\ HF^{n-1} \end{bmatrix} = n, \quad \text{rank}\,[K\ FK\ \ldots.\ F^{n-1}\ K] = n \tag{2.3}$$

n *is then called the order of the process* $\{y_t\}$.

ASSUMPTION 1b: *The polynomial matrices* $A(z)$ *and* $B(z)$ *are left coprime. It can then be shown that* $\deg \det A(z) = n$, *the order of the process.*

DEFINITION 1a: *The set of all minimal triples* (H,F,K) *of order* n *will be denoted by* S_n.

DEFINITION 1b: *The set of all left coprime polynomial pairs* $(A(z),B(z))$ *with* $\deg \det A(z) = n$ *will be denoted by* S_n^*.

Eliminating x_t in (2.1) or premultiplying (2.2) by $A^{-1}(z)$ leads to a third representation for the process $\{y_t\}$:

$$y_t = \sum_{i=1}^{\infty} H_i\, e_{t-1} = \underline{H}\, E^t \tag{2.4}$$

where the $p \times p$ matrices H_i are called impulse response matrices (or Markov parameters). The infinite matrix $\underline{H}$ is defined as $\underline{H} = [H_0 H_1 H_2 \ldots]$ with $H_0 = I_p$ and $H(z) = \sum_{i=1}^{\infty} H_i z^i$ is analytic in $|z| \le 1$. The infinite column vector E^t is defined as

$$E^t = \begin{bmatrix} e_t \\ e_{t-1} \\ e_{t-2} \\ \vdots \end{bmatrix}$$

The impulse response matrices are related to the representations (2.1) and (2.2) as follows:

$$H_0 = I, \quad H_i = H\,F^{i-1}\,K, \quad i = 1,2,\ldots \tag{2.5a}$$

$$\sum_{i=0}^{\infty} H_i z^{-i} = A^{-1}(z)B(z) \tag{2.5b}$$

The impulse response representation (2.4) completely specifies the second-order statistics of the process $\{y_t\}$, namely the covariance function $R_y(k) = E\,\{y_t y_{t-k}\}$, $k = 0,1,\ldots$ However there are several pairs $\{\underline{H},E^t\}$

which generate the same process $\{y_t\}$ (more specifically, a process $\{y_t\}$ with the same covariance kernel $R_y(k)$). To overcome this difficulty we require $\underline{H}$ to be not only causal, but also causally invertible, such that $\{e_t\}$ can be reconstructed from the past $\{y_t\}$ through a linear functional:

$$e_t = \sum_{i=0}^{\infty} G_i y_{t-i} = \underline{G}\, Y^t \quad , \tag{2.6}$$

with $G_0 = I_p$ and $G(z) = \sum_{i=0}^{\infty} G_i z^i$ is analytic in $|z| < 1$. $\underline{G}$ and Y^t are defined as $\underline{H}$ and E^t above.

Then $\{e_t\}$ is called the innovation of $\{y_t\}$ and

$$e_t \triangleq y_t - \hat{y}_{t/t-1} \tag{2.7}$$

where $\hat{y}_{t/t-1}$ is the linear least squares predictor of y_t given the past history Y^{t-1} of $\{y_t\}$.

We can now define identifiability up to second order statistics. Let θ be the vector of parameters in either the triple (H,F,K) or the pair $(A(z),B(z))$ and let Q be the covariance matrix of the white noise sequence $\{e_t\}$ in either (2.1) or (2.2).

DEFINITION 2: *2 parameter pairs* (θ_1,Q_1) *and* (θ_2,Q_2) *are undistinguishable if and only if*

$$R_y(k;\theta_1,Q_1) = R_y(k;\theta_2,Q_2) \qquad \forall k \geq 0 \tag{2.8}$$

where $R_y(k;\theta_i,Q_i)$ *is the covariance function of the process* $\{y_t\}$ *generated by model* i.

Note that if $\{y_t\}$ is Gaussian (or if a second order identification method is used such as a prediction error method), then the probability law (or the loss function) is completely determined by the second order moments, and Definition 2 can be replaced by the following.

DEFINITION 2': *For a Gaussian process 2 parameter pairs* (θ_1,Q_1) *and* (θ_2,Q_2) *are undistinguishable iff*

$$p(Y_0^T;\theta_1,Q_1) = p(Y_0^T;\theta_2,Q_2) \qquad \forall Y_0^T \text{ and } \forall T > 0 \quad . \tag{2.9}$$

Now it is easy to show that (θ_1,Q_1) and (θ_2,Q_2) are indistinguishable iff

$$Q_1 = Q_2 \text{ and } H_i(\theta_1) = H_i(\theta_2), \quad i = 0,1,2,\ldots \tag{2.10}$$

Because $Q_1 = Q_2$, we shall in the sequel drop the explicit dependence of $R_y(k)$ or $p(Y_0^T)$ on Q. The indistinguishability concept induces an equivalence relation on the sets S_n and S_n^*, which we shall denote by the symbol $\sim$. It follows from (2.5) and (2.10) that

$$\theta_1 \sim \theta_2 \Leftrightarrow H_i(\theta_1) = H_i(\theta_2) \quad \forall i \tag{2.11}$$

$$\Leftrightarrow H_2 = H_1 T, \quad F_2 = T^{-1} F_1 T, \quad K_2 = T^{-1} K_1$$

$$\text{for some nonsingular matrix } T. \tag{2.12}$$

$$\Leftrightarrow A_2(z) = M(z)A_1(z), \quad B_2(z) = M(z)B_1(z) \tag{2.13}$$

$$\text{for some unimodular matrix } M(z).^* \tag{2.13}$$

Two matrix triples (H_1,F_1,K_1) and (H_2,F_2,K_2) (resp. two polynomial matrix pairs $(A_1(z),B_1(z))$ and $(A_2(z),B_2(z))$ are called *equivalent* if the relations (2.12) (resp. (2.13)) hold.

The covariance function (or the probability law in the Gaussian case) of the process $\{y_t\}$ is completely determined by specifying (H,F,K,Q) or $(A(z),B(z),Q)$. But because of the nonuniqueness induced by (2.11)-(2.12), in order to achieve identifiability, we have to find a *reparametrization* of the family $R_y(k;\theta)$ or $P(Y_0^T;\theta)$ in such a way that two different sets of parameters (in the reparametrized set) correspond to two different sequences of Markov parameters.

From now on we shall, for simplicity, assume that the process $\{y_t\}$ is Gaussian; identifiability is then defined by Definition 2'. All statements hold, up to second order statistics, for non-Gaussian processes if $p(Y_0^T;\theta)$ is replaced by $R_y(k;\theta)$.

What is needed to achieve identifiability is a factorization of the map $p : \theta \to (\cdot;\theta)$ in the following way:

$$\begin{array}{ccc} \mathfrak{S}_n & \xrightarrow{\quad p \quad} & P \\ & {\scriptstyle f}\searrow \quad \nearrow{\scriptstyle \hat{p}} & \\ & X_n & \end{array} \tag{2.14}$$

Here $\mathfrak{S}_n$ is either S_n or S_n^* (see Definitions 1 above); p is the map defined by the probability law; P is the image of p. The set X_n and the functions $f : \mathfrak{S}_n \to X_n$ and $\hat{p} : X_n \to P$ must satisfy the following conditions:

---

*A polynomial matrix $M(z)$ is unimodular if it is square and if $\det M(z)$ = non-zero constant.

a) for each $\theta \in \mathfrak{S}_n$, $\xi = f(\theta)$ is finite-dimensional (2.15a)

b) $p(\cdot;\theta) = \hat{p}(\cdot;f(\theta))$ for all $\theta \in \mathfrak{S}_n$ (2.15b)

c) $\hat{p}(\cdot;\xi_1) = p(\cdot;\xi_2) \Rightarrow \xi_1 = \xi_2$. (2.15c)

The function f consists of a finite number of scalar components, say $f_1,\ldots,f_k$, which form a complete system of invariants (see [5]) for the equivalence relation (2.11), since by b) and c):

$$\theta_1 \sim \theta_2 \Leftrightarrow \hat{p}(\cdot;f(\theta_1)) = \hat{p}(\cdot;f(\theta_2)) \Leftrightarrow f(\theta_1) = f(\theta_2) \quad . \tag{2.16}$$

The set X_n can be identified with the quotient sets $S_n/\sim$ or $S_n^*/\sim$, or, equivalently, with the class of all impulse response sequences $\underline{H}$ admitting a minimal realization of order n. Now it can be shown that, when $p > 1$, no single parametrization is able to describe all n-th order systems. Rather X_n can be described by a cover of local coordinates, or equivalently by a family of $\binom{n-1}{p-1}$ overlapping parametrizations of dimension $2np$. Each set of $2np$ invariants (i.e. each local parametrization) is defined by specifying p integer valued numbers $n_1,\ldots,n_p$ called "structure indices":

$$f_{n_1,\ldots,n_p} : \mathfrak{S}_n^{(n_1,\ldots,n_p)} \rightarrow X_n, \quad \text{with} \quad S_n \subset \mathbf{R}^{2np} \tag{2.17}$$

where $\mathfrak{S}_n(n_1,\ldots,n_p)$ is a subset of $\mathfrak{S}_n$.

Each map (2.17) is locally a complete system of surjective invariants of dimension $2np$; the subsets $\mathfrak{S}_n(n_1,\ldots,n_p)$ for all possible choices of $n_1,\ldots,n_p$ overlap, and cover $\mathfrak{S}_n$.

In the next section we shall define the structure indices and show how a choice of structure indices $n_1,\ldots,n_p$ defines a set of $2np$ invariants $f_{n_1,\ldots,n_p}$. These $2np$ invariants are computable functions of the impulse response matrices $H_0,H_1,H_2,\ldots$ and will be called intrinsic invariants. We shall then show how to construct overlapping state-space or ARMA parametrizations as a function of the $2np$ intrinsic invariants.

3. CONSTRUCTION OF A COMPLETE SYSTEM OF INVARIANTS. From (2.4) and (2.7) we can write the linear least squares k-step ahead predictor $\hat{y}_{t+k|t}$ as follows:

$$\hat{y}_{t+k|t} = \sum_{i=k}^{\infty} H_i e_{t-i} \qquad k = 0,1,2,\ldots \quad . \tag{3.1}$$

Therefore

$$\hat{Y}_t \triangleq \begin{bmatrix} \hat{y}_{t+1|t} \\ \hat{y}_{t+2|t} \\ \hat{y}_{t+3|t} \\ \vdots \end{bmatrix} = \begin{bmatrix} H_1 & H_2 & H_3 & \cdots \\ H_2 & H_3 & H_4 & \cdots \\ H_3 & H_4 & H_5 & \cdots \\ \vdots & & & \end{bmatrix} \begin{bmatrix} e_t \\ e_{t-1} \\ e_{t-2} \\ \vdots \end{bmatrix} = \mathfrak{H} E^t \tag{3.2}$$

Since the process is of order n, the rank of the Hankel matrix $\mathfrak{H}$ is n. Therefore we can choose a set of linearly independent rows of $\mathfrak{H}$, which will form a basis for the whole row space of $\mathfrak{H}$. Note that the corresponding components of $\hat{Y}_t$ will then form a basis for the space spanned by all components of the prediction vector $\hat{Y}_t$. Now the structure indices will define which rows of $\mathfrak{H}$ will form the basis, and we shall show how to construct a set of $2np$ invariants from $\mathfrak{H}$ for a given choice of structure indices.

To any choice of n linearly independent rows of $\mathfrak{H}$ we shall associate a *multiindex* $\underline{i} = (i_1,\ldots,i_n)$ where the numbers $i_1,\ldots,i_n$, arranged in increasing order, are the indices of the rows of $\mathfrak{H}$ that form the basis. Two restrictive conditions will be imposed on the selection of the basis rows:

CONDITION 1: *if* $j \in \underline{i}$, *then* $j-p \in \underline{i}$

CONDITION 2: $1,2,\ldots,p \in \underline{i}$

Condition 1 follows from the structure of the Hankel matrix: if the (j-p)-th row of $\mathfrak{H}$ is in the span of the preceding rows, so is the j-th row. Condition 2 results from the full rank assumption on $\{y_t\}$: it follows that the p components of $\hat{y}_{t+1|t}$ are linearly independent.

DEFINITION 3: *If the selection of the basis vectors obeys conditions* 1 *and* 2, *then the corresponding multiindex is called "nice."*

All nice multiindices correspond to a choice of the basis inside the first $n-p+1$ row blocks of $\mathfrak{H}$. For given n and p, there are only a finite number of possible nice multiindices (Example: for a 2-dimensional vector process $(p=2)$ or order 3 $(n=3)$, there are only 2 possible nice multiindices: $\underline{i}_1 = (1,2,3)$ and $\underline{i}_2 = (1,2,4)$). For most n-th order processes of dimension p, several possible choices exist for the selection of a "nice multiindex" basis. On the other hand, there are subsets of $\mathfrak{S}_n$ for which only one basis exists.

DEFINITION OF THE STRUCTURE INDICES: *Let* $\underline{i} = (i_1,\ldots,i_n)$ *be a nice multiindex defining a basis for the rows of* $\mathfrak{H}$. *For* $k=1,\ldots,p$, *let* n_k *be the least natural number such that* $(k+n_k p) \notin \underline{i}$. *Then* $n_1,\ldots,n_p$ *are called the "structure indices" corresponding to the basis; they specify which rows of* $\mathfrak{H}$ *are taken in the basis. Note that* $\sum_{i=1}^{p} n_i = n$.

We can now define $\mathfrak{S}_n(n_1,\ldots,n_p)$ as the set of all n-th order systems for which the n rows specified by $n_1,\ldots,n_p$ in the Hankel matrix are linearly independent. $\mathfrak{S}_n(n_1,\ldots,n_p)$ is a proper subset of $\mathfrak{S}_n$.

Consider now an element of $\mathfrak{S}_n(n_1,\ldots,n_p)$ specified by its Hankel matrix $\mathfrak{H}$. We shall construct a complete system of 2np surjective invariants for this system, i.e. a reparametrization of this system using 2np parameters.

Let H^i be the i-th block of p rows of the infinite Hankel matrix $\mathfrak{H}$ (e.g., $H^2 = [H_2 H_3 H_4 \ldots]$ and let

$$H^i = \begin{bmatrix} h_{1i} \\ h_{2i} \\ \vdots \\ h_{pi} \end{bmatrix} \tag{3.3}$$

where h_{ki} are rows of infinite length. Since $\underline{H}$ is an element of $\mathfrak{S}_n(n_1,\ldots,n_p)$, the rows $(h_{11},\ldots,h_{1n_1}; h_{21},\ldots,h_{2n_2}; \ldots; h_{p1},\ldots,h_{pn_p})$ form a basis. Therefore the rows $h_{1(n_1+1)}, h_{2(n_2+1)},\ldots, h_{p(n_p+1)}$ can be expressed as:

$$h_{i(n_{i+1})} = \sum_{j=1}^{p} \sum_{k=1}^{n_j} \alpha_{ijk} h_{jk} \qquad i = 1,\ldots,p \tag{3.4}$$

These relations define np scalar numbers α_{ijk}.

Now denote by $h_{ij}(k)$ the element in row i, column j of H_k. Then the 2np numbers

$$\{\alpha_{ijk},\ k = 1,\ldots,n_j;\ h_{ij}(k),\ k = 1,\ldots,n_i;\ i,\ j = 1,\ldots,p\} \tag{3.5}$$

completely coordinatize $\mathfrak{S}_n(n_1,\ldots,n_p)$, i.e. they map that set in a one to one manner, on Euclidean space of dimension 2np. The impulse response sequence $H_1, H_2, H_3,\ldots$ is completely specified by the p structure indices and these 2np numbers. These 2np numbers constitute a complete system of surjective invariants, which will be called "intrinsic invariants" of the process. In the notation of Section II:

$$f_{n_1,\ldots,n_p} : \mathcal{S}_n(n_1,\ldots,n_p) \to \mathbb{R}^{2np}$$

$$: \{H_1,H_2,H_3,\ldots\} \to f_{n_1,\ldots,n_p} = \{\alpha_{ijk},h_{ij}(k)\} \qquad (3.6)$$

The word intrinsic is used because the invariants are defined from the (intrinsic) infinite impulse response representation, and not from a finite state-space or ARMA model.

We have thus constructed a family of cuntions $f_{n_1,\ldots,n_p}$, each of which is a complete system of surjective invariants mapping onto $\mathbb{R}^{2np}$. These functions are defined on the overlapping subsets $\mathcal{S}_n(n_1,\ldots,n_p)$ which cover $\mathcal{S}_n$. Next we show that one can construct corresponding overlapping (state-space or ARMA) parametrizations whose parameters are functions of the $2np$ intrinsic invariants just defined.

STATE-SPACE PARAMETRIZATION

Consider an element of $\mathcal{S}_n(n_1,\ldots,n_p)$ for a given set of structure indices, and let $\{\alpha_{ijk},h_{ij}(k)\}$ be the instrinsic invariants of that element. Then the following is a state-space representation of that element:

$$H = \left.\begin{bmatrix} \underbrace{\begin{matrix} 1 & 0 & \ldots & 0 \\ 0 & . & & . \\ . & . & & . \\ . & . & & . \\ . & . & & . \\ 0 & 0 & \ldots & 0 \end{matrix}}_{n_1} & \Bigg| & \underbrace{\begin{matrix} 0 & 0 & \ldots & 0 \\ 1 & 0 & \ldots & 0 \\ 0 & . & & . \\ . & . & & . \\ . & . & & . \\ 0 & . & \ldots & 0 \end{matrix}}_{n_2} & \Bigg| & \ldots\ldots & \Bigg| & \underbrace{\begin{matrix} 0 & 0 & \ldots & 0 \\ . & . & & . \\ . & . & & . \\ . & . & & . \\ 0 & 0 & & . \\ 1 & 0 & \ldots & 0 \end{matrix}}_{n_p} \end{bmatrix}\right\} P \qquad (3.7a)$$

$$F = \left[\begin{array}{cccc|cccc|c|cccc}
0 & & & & 0 & \cdots & & 0 & & 0 & & & 0 \\
\vdots & & I_{n_1-1} & & \vdots & & & \vdots & \cdots & \vdots & & & \vdots \\
0 & & & & 0 & \cdots & & 0 & & 0 & & & 0 \\
\alpha_{111} & \cdots & & \alpha_{11n_1} & \alpha_{121} & \cdots & & \alpha_{12n_2} & & \alpha_{1p1} & \cdots & & \alpha_{1pn_p} \\
\hline
& \vdots & & & & \vdots & & & & & \vdots & & \\
\hline
0 & \cdots & & 0 & 0 & \cdots & & 0 & & 0 & & & \\
\vdots & & & \vdots & \vdots & & & \vdots & \cdots & \vdots & & I_{n_p-1} & \\
0 & \cdots & & 0 & 0 & \cdots & & 0 & & 0 & & & \\
\alpha_{p11} & \cdots & & \alpha_{p1n_1} & \alpha_{p21} & \cdots & & \alpha_{p2n_2} & & \alpha_{pp1} & \cdots & & \alpha_{ppn_p}
\end{array}\right] \tag{3.7b}$$

$$K = \begin{vmatrix} K_1 \\ K_2 \\ \vdots \\ K_p \end{vmatrix}, \quad \text{with} \quad K_i = \begin{vmatrix} k_{i1} \\ \vdots \\ k_{in_i} \end{vmatrix}, \quad \text{and} \quad k_{ij} = [h_{i1}(j), \ldots, h_{ip}(j)],$$

a row p-vector (3.7c)

The proof is a straightforward but tedious verification that the relations (2.5a) hold with the α_{ijk} and $h_{ij}(k)$ defined from the H_i as in (3.4)-(3.5)

The process $\{y_t\}$, which is by assumption an element of $\mathfrak{S}_n(n_1, \ldots, n_p)$, can be represented by (2.1) with H, F, K as in (3.7).

ARMA PARAMETRIZATION

By an argument similar to that developed in [5] and [13] we obtain the following equations for the entries $A(z)$ and $B(z)$ of (2.2b):

$$a_{ii}(z) = z^{n_i} - \alpha_{iin_i} z^{n_i - 1} - \ldots - \alpha_{ii1} \tag{3.12a}$$

$$a_{ij}(z) = - \alpha_{ijn_j} z^{n_j - 1} - \ldots - \alpha_{ij1} \tag{3.12b}$$

and

$$b_{ij}(z) = b_{ij\bar{n}+1} z^{n} + b_{ij\bar{n}}\, z^{\bar{n}-1} + \dots + b_{ij1} \tag{3.12c}$$

with $\bar{n} = \max_{1\le i\le p} n_i$.

The α_{ijk} are defined by (3.4), while the b_{ijk} are defined as follows

$$\bar{B} = M\,\bar{K} \tag{3.13}$$

where

$$\underset{(\bar{n}+1)p\times p}{\bar{B}} = \begin{bmatrix} B^1 \\ \vdots \\ B^p \end{bmatrix}, \qquad \underset{(n+1)\times p}{B^i} = \begin{bmatrix} b_{i11} & & b_{ip1} \\ \vdots & & \\ b_{i1(n+1)} & \cdots & b_{ip(n+1)} \end{bmatrix} \tag{3.14}$$

$$\underset{(n+p)\times p}{\bar{K}} = \begin{bmatrix} (1 \;\; 0 \; \dots \; 0) \\ k_{11} \\ \cdot \\ \cdot \\ \cdot \\ k_{1n_1} \\ (0 \;\; 1 \;\; 0 \dots 0) \\ k_{21} \\ \cdot \\ \cdot \\ \cdot \\ k_{pn_p} \end{bmatrix} \tag{3.15}$$

with k_{ij} defined by (3.7c)

$$M = [M_{ij}] \qquad (i,j = 1,\dots,p)$$

with

$$\underset{(n+1)\times(n_i+1)}{M_{ii}} = \begin{array}{r} \\ \\ \\ \\ \\ \\ \bar{n}-n_i \left\{ \begin{array}{c} \\ \\ \end{array}\right. \end{array} \begin{bmatrix} -\alpha_{ii1} & \cdots\cdots & -\alpha_{iin_i} & 1 \\ & & \cdot & \\ & \cdot & & \\ -\alpha_{iin_i} & \cdot & & \\ 1 & \cdots\cdots & & \\ 0 & \cdots\cdots & & 0 \\ 0 & \cdots\cdots & & 0 \end{bmatrix} \tag{3.16}$$

and

$$\underset{(n+1)\times(n_j+1)}{M_{ij}} = \begin{array}{r} \\ \\ \\ \\ \\ \\ \bar{n}-n_j \left\{ \begin{array}{c} \\ \\ \end{array}\right. \end{array} \begin{bmatrix} -\alpha_{ij1} & \cdots\cdots & -\alpha_{ijn_j} & 0 \\ & & \cdot & \\ & \cdot & & \\ -\alpha_{ijn_j} & \cdot & & \\ 0 & \cdots\cdots & & \\ 0 & \cdots\cdots & & 0 \\ 0 & \cdots\cdots & & 0 \end{bmatrix} \tag{3.17}$$

We can now summarize the main points of this section by extending the scheme (2.14). For every element of $\mathfrak{S}_n(n_1,\ldots,n_p)$, we have first defined the intrisic invariants $f_{n_1,\ldots,n_p}$. Next, by (3.7), (3.12) and (3.13), we have established two bijections $g_1(n_1,\ldots,n_p)$ and $g_2(n_1,\ldots,n_p)$ between the intrinsic invariants (defined on $\underline{H}$) and the quotient spaces $S_n/_\sim$ and $S_n^*/_\sim$

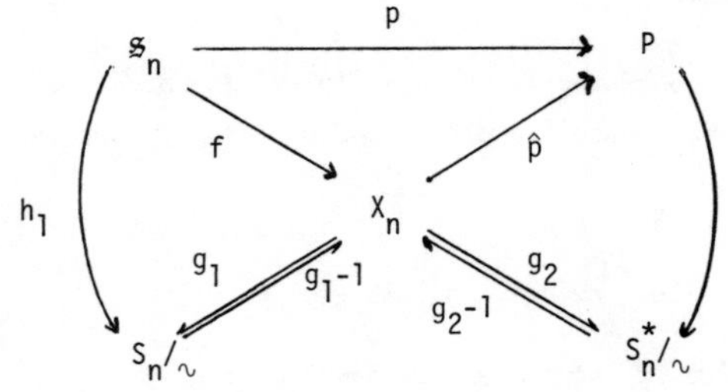

(3.18)

Hence, by a property of complete sets of surjective invariants (see, e.g., [5]), $h_i(n_1,\ldots,n_p) = g_i(n_1,\ldots,n_p) \circ f_{n_1\ldots n_p}$ are also locally complete sets of surjective invariants. This leads to the following factorizations:

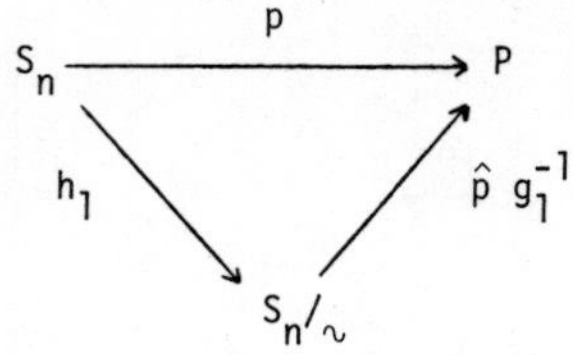

(3.19a)

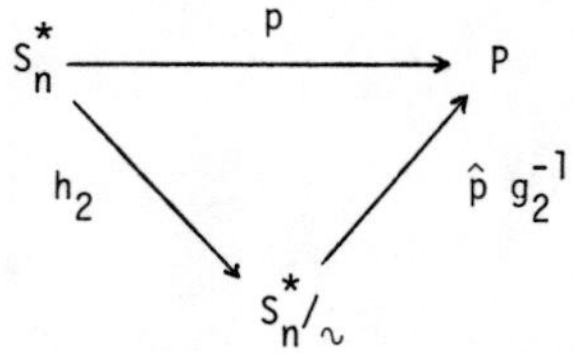

(3.19b)

In figs. (3.18) - (3.19) all quantities are to be indexed by $(n_1,\ldots,n_p)$. These factorizations of the probability map make the multivariable models identifiable. Given an arbitrary element of $S_n(n_1,\ldots,n_p)$ (i.e. a matrix triple H, F, K for which the rows of $\mathfrak{H}$ specified by the structure indices $n_1,\ldots,n_p$ form a basis) which is parametrized by $n^2 + 2np$ parameters that are not identifiable, we replace this element by an equivalent element of the quotient space $S_n/_\sim$ via the map $h_1(n_1,\ldots,n_p)$ (see Fig. 3.19a). This last element is parametrized by the system of $2np$ invariants defined by h_1 and appearing in the form (3.7). These $2np$ invariants are uniquely identifiable. The same can be said if an ARMA form is used (see Fig. 3.19b).

4. ASYMPTOTIC EQUIVALENCE OF ALL OVERLAPPING FORMS. In most cases an n-th order system can be represented in more than one of the overlapping forms. This corresponds to the fact that in general different choices of nice multi-indices can be made, which will define different sets of linearly independent rows of the Hankel matrix.

EXAMPLE. *Consider a bivariate process* $(p = 2)$ *or order* 3 $(n = 3)$. *Two nice multiindices exist, with the corresponding sets of structure indices:*

$$i_1 = (1,2,3), \quad \text{i.e.} \quad n_1 = 2, \quad n_2 = 1$$

$$i_2 = (1,2,4), \quad \text{i.e.} \quad n_1 = 1, \quad n_2 = 2 \quad .$$

The corresponding sets of intrinsic invariants are:

- for $i_1 : \alpha_{111}, \alpha_{112}, \alpha_{121}, \alpha_{211}, \alpha_{212}, \alpha_{221}$

  $h_{11}(1),\ h_{12}(1),\ h_{21}(1),\ h_{22}(1),\ h_{11}(2),\ h_{12}(2).$

- for $i_2 : \alpha_{111}, \alpha_{121}, \alpha_{122}, \alpha_{211}, \alpha_{221}, \alpha_{222}$,

$$h_{11}(1),\ h_{12}(1),\ h_{21}(1),\ h_{22}(1),\ h_{21}(2),\ h_{22}(2).$$

The corresponding state-space parameterizations are:

$$H_1 = \begin{bmatrix} 1 & 0 & 0 \\ 0 & 0 & 1 \end{bmatrix} \quad F_1 = \begin{bmatrix} 0 & 1 & 0 \\ \alpha_{111} & \alpha_{112} & \alpha_{121} \\ \alpha_{211} & \alpha_{212} & \alpha_{221} \end{bmatrix} \quad K_1 = \begin{bmatrix} h_{11}(1) & h_{12}(1) \\ h_{11}(2) & h_{12}(2) \\ h_{21}(1) & h_{22}(1) \end{bmatrix}$$

$$H_2 = \begin{bmatrix} 1 & 0 & 0 \\ 0 & 1 & 0 \end{bmatrix} \quad F_2 = \begin{bmatrix} \alpha_{111} & \alpha_{121} & \alpha_{122} \\ 0 & 0 & 1 \\ \alpha_{211} & \alpha_{221} & \alpha_{222} \end{bmatrix} \quad K_2 = \begin{bmatrix} h_{11}(1) & h_{12}(1) \\ h_{21}(1) & h_{22}(1) \\ h_{21}(2) & h_{22}(2) \end{bmatrix}$$

Now it can be shown (see e.g. [7]-[9]) that the state x_t of the state-space realization (2.1), with H, F, K defined by (3.7), is made up of the components $\hat{Y}_{t-1}$ indexed by the elements of the selected nice multiindex. In our example, the state is defined in one of two possible ways according to which choice of structure indices is made:

for $n_1 = 2,\ n_2 = 1$ — for $n_1 = 1,\ n_2 = 2$

$i_1 = (1,2,3)$ — $i_2 = (1,2,4)$

$$x_t \triangleq \begin{bmatrix} \hat{y}^1_{t/t-1} \\ \hat{y}^1_{t+1/t-1} \\ \hat{y}^2_{t/t-1} \end{bmatrix} \qquad x_t \triangleq \begin{bmatrix} \hat{y}^1_{t/t-1} \\ \hat{y}^2_{t/t-1} \\ \hat{y}^2_{t+1/t-1} \end{bmatrix}$$

The structure of the corresponding H, F, K matrices is shown above. In most cases, both choices are possible, because both sets of 3 components form a linearly independent set. The question then is whether one choice is better than the other. For example, one might think that if $\hat{y}^1_{t+1/t-1}$ is close to the linear span of $\hat{y}^1_{t/t-1}$ and $\hat{y}^2_{t/t-1}$, then the choice of i_2 would be preferable because the components of the state would be more orthogonal to one another, thereby making the ensuing parameter estimation problem numerically better behaved. More generally the question that has intrigued several researchers in recent years is whether any one of the overlapping parametrizations is optimal in the sense that, in that particular parametrization, the parameter estimates can be estimated with a higher accuracy or that the estimation algorithm will have a better numerical behaviour. In the main result of this paper we present a partial answer to this question.

THEOREM. *Given the intrinsic invariants* $\{\alpha_{ijk}, h_{ij}(k)\}$ and $\{\alpha^*_{ijk}, h^*_{ij}(k)\}$ *corresponding to two different sets of structure indices* $\{n_1, \ldots, n_p\}$ *and* $\{n^*_1, \ldots, n^*_p\}$, *then the determinants of the information matrices corresponding to these two parametrizations are identical.*

The proof of the Theorem follows from the following lemma.

LEMMA. *Let* $\{\alpha_{ijk}, h_{ij}(k)\}$ *and* $\{\alpha^*_{ijk}, h^*_{ij}(k)\}$ *be the intrinsic invariants of a given* n-*th order system in* $\mathcal{S}_n$ *for two different choices of the structure indices. Then the Jacobian of the transformation between these two parameter vectors is unity.*

The proof of the lemma can be found in [9]. The theorem can then be proved as follows. Let $\theta = \{\alpha_{ijk}, h_{ij}(k)\}$ and $\theta^* = \{\alpha^*_{ijk}, h^*_{ij}(k)\}$. The corresponding information matrices M_θ and M^*_θ are related by[†]

$$M_{\theta^*} = E_{Y|\theta^*}\left\{\left(\frac{\partial \log p(Y|\theta^*)}{\partial \theta^*}\right)^T\left(\frac{\partial \log p(Y|\theta^*)}{\partial \theta^*}\right)\right\}$$

$$= E_{Y|\theta^*}\left\{\left(\frac{\partial \log p(Y|\theta)}{\partial \theta}\ \frac{\partial \theta}{\partial \theta^*}\right)^T\left(\frac{\partial \log p(y|\theta)}{\partial \theta}\ \frac{\partial \theta}{\partial \theta^*}\right)\right\}$$

$$= \left(\frac{\partial \theta}{\partial \theta^*}\right)^T E_{Y|\theta}\left\{\left(\frac{\partial \log p(Y|\theta)}{\partial \theta}\right)^T\left(\frac{\partial \log p(Y|\theta)}{\partial \theta}\right)\right\}\left(\frac{\partial \theta}{\partial \theta^*}\right)$$

$$= \left(\frac{\partial \theta}{\partial \theta^*}\right)^T M_\theta \left(\frac{\partial \theta}{\partial \theta^*}\right) \quad .$$

It follows from the bijective relationship between the intrinsic invariants $\{\alpha_{ijk}, h_{ij}(k)\}$ and the corresponding overlapping parametrizations H, F, K or $A(z)$, $C(z)$ that the theorem also holds when two overlapping (state-space or ARMA) parametrizations are compared.

COROLLARY. *Given to overlapping parametrizations* F, K, H *and* F^*, K^*, H^* *in the form* (3.7) *(resp.* $A(z)$, $B(z)$ *and* $A^*(z)$, $B^*(z)$ *in the form* (3.12)) *for the same process, corresponding to two different sets of structure indices*

[†] If x is a scalar and θ a column k-vector, then $\frac{\partial x}{\partial \theta}$ denotes the row vector $\left[\frac{\partial x}{\partial \theta}, \ldots, \frac{\partial x}{\partial \theta_k}\right]$.

$\{n_1,\ldots,n_p\}$ and $\{n_1^*,\ldots,n_p^*\}$, then the determinants of the information matrices corresponding to these two parametrizations are identical.

If the parameters are estimated using a maximum likelihood or a prediction error method, then the covariance matrix of the estimation errors is asymptotically equal to the inverse of the Fisher information matrix M_θ. Therefore it follows from our main result that all overlapping parametrizations are asymptotically equivalent, as far as the accuracy of the parameter estimates is concerned, when this accuracy is measured by the determinant of the covariance matrix of the estimation errors. As a consequence this criterion is unable to discriminate between two overlapping structures for the same process. Of course other criteria would be used that might be able to discriminate, even asymptotically, between different structures, see e.g. [10].

Some structures might also be better than others when only a finite data record is available. In the next two sections we describe two heuristic basis selection procedures that can be used for the finite data case. They are based on the idea of selecting the "most independent components" in the state vector.

5. A METHOD BASED ON THE CONCEPT OF COMPLEXITY. In this section, we sketch a method proposed by Ljung and Rissanen [7] and based on the concept of complexity of a random vector, defined by Van Emden [14]. We shall also propose a new iterative procedure that is closely related to that of Ljung and Rissanen.

Complexity is in fact a measure of the interaction between the components of a random vector. The more interaction there is, the larger the complexity. Van Emden shows that the complexity can be expressed using the covariance matrix and derives the following expression:

$$C = -\frac{1}{2}\sum_{i=1}^{n} \log(n\lambda_i) \tag{5.1}$$

where λ_i are the eigenvalues of the covariance matrix of the random vector (provided this covariance matrix has been normalized so that its trace equals 1).

Suppose now that we know the covariance matrix $R_{\hat{Y}}$ of $\hat{Y}_t^N$ and that the order of the process, n, is also known. Then one can compute the complexity of various subvectors of order n of $\hat{Y}_t^N$, because the corresponding $n \times n$ covariance matrices are submatrices of $R_{\hat{Y}}$. The idea proposed in [7] is then to select as the state that subvector of $\hat{Y}_t^N$ that has the smallest complexith among all subvectors of dimension n that obey the conditions 1 and 2 of Section 3. The components obtained in this way are called by Ljung and Rissanen the "most independent components" of Y_t^N. The procedure they suggest is as follows:

(i) compute estimates of the predictors $\hat{y}_{t+k|t}$ by first fitting a high-order autoregressive model to the data.

(ii) compute the sample covariance matrix $R_{\hat{Y}}$ from the estimate predictions.

(iii) for a given value of n, compute the complexity of various submatrices of $R_{\hat{Y}}$, subject to the constraints that the $p \times p$ upper left submatrix of $R_{\hat{Y}}$ is always included and that the j-th row of the matrix $R_{\hat{Y}}$ is chosen only if the (j-p)-th row is also chosen. (These constraints amount to meeting conditions 1 and 2 of Section 3).

(iv) select the basis for the predictor space that corresponds to the submatrix with smallest complexity.

(v) repeat the procedure for higher order models and take the order that minimizes a criterion such as Akaike's AIC criterion.

A major disadvantage of this method is that one needs to first estimate the covariance matrix of the prediction vector by fitting a high-order autoregressive model to the data and then computing sample predictions. However the procedure has the following interesting feature: if the parameters of the F matrix are estimated by least squares, one can show (see [15]) that the covariance matrix of the error of the parameter estimates is related to the inverse of the submatrix of $R_{\hat{Y}}$ selected by the procedure of Ljung and Rissanen. This seems to justify using the matrix $R_{\hat{Y}}$ as a starting point for the selection of the basis components, even though this matrix is not directly available. It also suggests minimizing some scalar measure of the inverse of the various submatrices of $R_{\hat{Y}}$ in order to discriminate between the corresponding subvectors of $\hat{Y}_t^N$. With this idea in mind, we suggest the following procedure.

The first two steps are identical to those in Ljung and Rissanen's method.

(iii) compute the inverse of the upper left $p \times p$ submatrix of $R_{\hat{Y}}$.

(iv) for an order n equal to $p+1$, select all the $(p+1) \times (p+1)$ submatrices of $R_{\hat{Y}}$, which contain the $p \times p$ upper left submatrix, and such that condition 1 of Section 3 is also satisfied. Compute the inverses of these submatrices, using the fact that the inverse of a matrix $A^1 = \begin{bmatrix} A & b \\ b^T & \alpha \end{bmatrix}$, where b is a vector and α a scalar, is given by the following inversion formulae:

$$(A^1)^{-1} = \begin{bmatrix} E & g \\ g^T & \varepsilon \end{bmatrix}$$

where

$$E = A^{-1} + A^{-1}b\varepsilon b^T A^{-1}$$

$$g = -A^{-1}b\varepsilon$$

$$\varepsilon = [\alpha - b^T A^{-1} b]^{-1} \quad .$$

(Hence, no other matrix inversion is needed once the $p \times p$ submatrix A^{-1} has been computed.)

(v) Select the submatrix for which the trace of the inverse is minimized.

(vi) Repeat the last two steps with all $(n+1) \times (n+1)$ submatrices that contain the selected $n \times n$ submatrix and whose additional row and column is chosen so as to satisfy condition 1 of Section 3.

(vii) A stopping criterion is needed. One criterion that seems to work well in practice is to stop when $J = \frac{1}{n} \operatorname{tr} [(A^1)^{-1}]$ does not decrease anymore. This is justified by the fact that when the order of the submatrices is greater than the actual order of the process, these submatrices will be ill-conditioned. Their inverses will be large and so will be J.

The method described in this Section is an off-line method: a "best basis" is chosen a priori, and the parameters are subsequently estimated in that basis. Van Overbeek and Ljung [8] have proposed an alternative on-line procedure, which is also derived from [7]. In their scheme, the parameter estimation procedure starts with any parametrization that satisfies conditions 1 and 2 of Section 3. During the estimation procedure, the covariance matrix of the state of the representation is monitored; if that matrix becomes ill-conditioned a similarity transformation is applied in such a way that the new coordinate system corresponds to a better conditioned basis.

6. A Q-R FACTORIZATION METHOD. In this section, we present a new method of structure identification, based on the assumption that a "good" structure will be one in which the predictors selected to form the basis of the prediction space are most independent. We will first assume, as has been done by Akaike [3], that we have chosen $M \in \mathbb{N}$ large enough so that $\underline{H}(Y^t_{t-M})$ (i.e. the Hilbert space spanned by the components of $y(t-k)$ for $0 \le k \le M$) is close enough to the sapce $\underline{H}(Y^t_{-\infty})$. This allows us to replace the analysis of dependence of the $\hat{y}^j_{t+k|t}$ by the analysis of dependence of the $\hat{y}^j_{t+k|t,t-M}$, which are defined as the projections of y^j_{t+k} onto the space $\underline{H}(Y^t_{t-M})$.

Denote by Y^t_{t-M} and Y^t_{t+N} the vectors

$$Y^t_{t-M} = \begin{bmatrix} y(t) \\ y(t-1) \\ y(t-M) \end{bmatrix} \qquad Y^t_{t+N} = \begin{bmatrix} y(t) \\ y(t+1) \\ y(t+N) \end{bmatrix} \tag{6.1}$$

and by Σ_{11}, Σ_{12} and Σ_{22} the covariance and cross covariance matrices

$$\Sigma_{11} = E\{Y^t_{t+N}(Y^t_{t+N})^T\} , \quad \Sigma_{12} = E\{Y^t_{t+N}(Y^t_{t-M})^T\}$$

$$\Sigma_{22} = E\{Y^t_{t-M}(Y^t_{t-M})^T\} \quad . \tag{6.2}$$

By the full rank assumption Σ_{22} is positive definite so that we can have

$$\Sigma_{22}^{-T/2}\Sigma_{22}\Sigma_{22}^{-1/2} = I \tag{6.3}$$

for some nonsingular matrix $\Sigma_{22}^{1/2}$.

Let us perform the following transformation:

$$Z^t_{t-M} = \Sigma_{22}^{-T/2}\ Y^t_{t-M} \tag{6.4}$$

so that

$$E\{(Z^t_{t-M})(Z^t_{t-M})^T\} = I \tag{6.5}$$

Now $\hat{y}^j_{t+k|t,t-M}$ admits a representation:

$$\hat{y}^j_{t+k|t,t-M} = \sum_{m=0}^{M}\sum_{\ell=1}^{p} a(j,k,\ell,m)z^\ell_{t-m} \tag{6.6}$$

where

$$E\{\hat{y}^j_{t+k|t,t-M}\ z^\ell_{t-m}\} = a(j,k,\ell,m) = E\{y^j_{t+k}\ z^\ell_{t-m}\} \quad . \tag{6.7}$$

Hence, the elements of the (kp+j)-th row of matrix $\Sigma_{12}\Sigma_{22}^{-1/2}$ are the coordinates of $\hat{y}^j_{t+k|t,t-M}$ in the basis Z^t_{t-M}. Therefore, because of (6.5), we can replace the analysis of independence of the components $\hat{y}^j_{t+k|t,t-M}$ by the analysis of independence of the rows of $\Sigma_{12}\Sigma_{22}^{-1/2}$. Now, a well known (and numerically well conditioned) method to search for the most independent rows of a given matrix is to perform a Q-R factorization by means of Householder transformations (see [16] and [17]) with row-interchange.

The method we propose is thus as follows:

(i) compute an upper-triangular square root of Σ_{22}, i.e. an uppertriangular matrix $\Sigma_{22}^{1/2}$ such that $\Sigma_{22}^{T/2}\Sigma_{22}^{1/2} = \Sigma_{22}$ (Cholesky factorization).

(ii) compute the product $\Sigma_{12}\Sigma_{22}^{-1/2}$. Note that the first p rows of this product are lower triangular. This saves p steps in the subsequent QR factorization procedure.

(iii) compute a recursive Q-R factorization of $\Sigma_{12}\Sigma_{22}^{-1/2}$ using Householder transformations, where the triangularization is always performed on the row leading to the largest pivot.

(iv) stop when the pivots do not significantly decrease any more.

We briefly illustrate one step of the Q-R factorization.

After k steps of the triangularization procedure, we have the following factorization:

$$S_k^T \Sigma_{12} \Sigma_{22}^{-1/2} P_k = T_k$$

where S_k is the product of k permutation matrices, P_k is the product of k Householder transformation matrices, and T_k is a lower triangular matrix of the following form:

$$T_k = \left[\begin{array}{cccccc|ccc} t_1 & 0 & & & & & 0 & \cdots & 0 \\ X & t_2 & 0 & \cdots & 0 & & \vdots & & \vdots \\ \vdots & & & & & \vdots & & & \\ \vdots & & & & & 0 & \vdots & & \vdots \\ X & \cdots & & X & & t_k & 0 & \cdots & 0 \\ \hline & & & & & & & & \\ & & T_{21} & & & & & T_{22} & \\ & & & & & & & & \end{array}\right]$$

It is clear from the structure of T_k that the Euclidean norm of the rows in T_{22} are the distances of the last $(N+1)p-k$ rows of T_k to the space spanned by the first k rows. From the properties of Householder transformations, the pivot of the next triangularization step is the Euclidean norm of the selected row in T_{22}. So choosing the row leading to the largest pivot amounts to choosing the row with the largest projection onto the space orthogonal to the span of the first k rows. This is what we call the "most independent" row.

Upon completion of the triangularization procedure, the following factorization is obtained:

$$\Sigma_{12} \Sigma_{22}^{-1/2} = STP^T \tag{6.8}$$

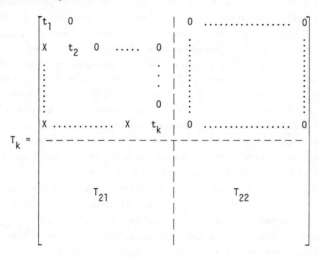

where S is a permutation matrix, T is a lower triagular matrix (at least for its first n rows) and P is a product of Householder transformation matrices (which implies that $P^TP = PP^T = I$). The permutation matrix S indicates which components of $\hat{Y}_t^N$ are to be chosen in the basis. In order to be consistent with the conditions 1 and 2 of Section 3, we should also introduce some constraints in the procedure of row selection for the triangularization: the first p rows are to be triangularized, and one can choose the j-th row only if the (j-p)-th row has already been chosen.

We now show that our triangularization method can also be related very nicely to Akaike's method [3] which is based on a canonical correlation analysis on the vectors of future and past observations (see Anderson [18]). We shall not go back in detail to this method, but we recall that it amounts to a singular value decomposition of the matrix $\Sigma_{11}^{-T/2}\Sigma_{12}\Sigma_{22}^{-1/2}$ and that the idea of the canonical correlation analysis in this case is to search for independent linear combinations of the vector Y_{t+N}^{t} which are most correlated with independent linear combinations of Y_{t-M}^{t}. Actually, in Akaike's method, this only gives the number of independent components of a sub-vector of the predictor vector, and hence it leads to the choise of the *first* independent components of the prediction vector in the basis. It is not possible with this method to discriminate between various independent components and to take the "most independent" ones, and the reason for this is that the canonical correlation analysis uses linear combinations of Y_{t+N}^{t} rather than single components of this vector. By searching for the *single* components of Y_{t+N}^{t} that are most correlated with independent linear combinations of the past, one can establish a close relationship between our procedure and the method of Akaike.

In the canonical correlation analysis, one searches in a first step for two vectors α_1 and γ_1 such that $\alpha_1^T\Sigma_{12}\gamma_1$ is maximized subject to $\alpha_1^T\Sigma_{11}\alpha_1 = \gamma_1^T\Sigma_{22}\gamma_1 = 1$. In a second step, two linear combinations $\alpha_2 Y_{t+N}^{t}$ and $\gamma_2 Y_{t-M}^{t}$ are sought, which are orthogonal to the first ones, and which have maximum correlation with one another. Now consider the following variations.

In the first step we maximize $a_1(\Sigma_{12})_{j_1}\gamma_1$, where a_1 is a scalar, $(\Sigma_{12})_{j_1}$ is the j_1-th row of Σ_{12}, γ_1 is a vector of dimension $(M+1)p$, under the constraints $a_1^2 = 1$, $\gamma_1^T\Sigma_{22}\gamma_1 = 1$. The maximum is taken over the values of a_1, γ_1 and the index j_1. For the second step, we maximize $a_2(\Sigma_{12})_{j_2}\gamma_2$, $j_2 \neq j_1$, under the constraints $a_2^2 = 1$, $\gamma_2^T\Sigma_{22}\gamma_2 = 1$, $\gamma_2^T\Sigma_{22}\gamma_1 = 0$, and so on.

With computations similar to those of the canonical correlation analysis, this amounts to finding two matrics A and Γ such that $A^TA = I$ and A is a permutation matrix (with possibly some changes of sign), $\Gamma^T\Sigma_{22}\Gamma = I$ and

$$A^T\Sigma_{12}\Gamma = T \tag{6.9}$$

T is a lower triangular matrix with decreasing pivots. If we compare (6.8) and (6.9), we can identify S with A and $\Sigma_{22}^{-1/2}P$ with Γ, and hence establish the equivalence between the two approaches.

In this section we have proposed a new structure estimation method as an alternative to the method of Rissanen and Ljung. It is based on another heuristic definition of "most independent rows" of a matrix. We do not claim that this new method is superior from a theoretical point of view. However, from a computational point of view, our method has the major advantage that it works with the covariance function of the observation process, R_Y, which can be readily estimated from the data, while the method of Rissanen and Ljung works with the covariance of the predictors, $R_{\hat{Y}}$, which requires that the data be filtered first, using an AR model that has to be identified.

7. SIMULATION RESULTS. In this section, we present some simulation results in order to compare the various methods that we have introduced. Data sequences have been generated from the Markovian model:

$$x_k = Ax_{k-1} + Be_k$$

$$y_k = Cx_k \quad .$$

A member of different models have been simulated but, for reasons of brevity, we restrict ourselves here to two models which will hopefully give a sufficient illustration for the previous sections. In both models, y_k is a two-dimensional vector process, x_k is of dimension 4, and e_k is a two-dimensional Gaussian white noise, with mean zero and unit covariance matrix.

Tables 1 and 2 gives the values of the matrices A, B and C for each of the two models

$$A_1 = \begin{bmatrix} -0.50 & 4.83 & -0.63 & 0.72 \\ 0.20 & -3.67 & 0.50 & -0.58 \\ -0.22 & -2.42 & -0.19 & -0.36 \\ -1.55 & 22.50 & -2.87 & 3.69 \end{bmatrix} \qquad B_1 = \begin{bmatrix} 3.38 & -3.13 \\ -2.5 & 2.5 \\ -0.69 & 1.56 \\ 15.13 & -14.88 \end{bmatrix}$$

$$C_1 = \begin{bmatrix} -70 & 38.0 & 16.0 & 9.0 \\ -4.39 & 85.33 & 37.5 & 17.92 \end{bmatrix}$$

Table 1: Parameter values for model 1

$$A_2 = \begin{bmatrix} 0 & 1 & 0 & 0 \\ -0.1 & 0.65 & 0 & 0 \\ 0 & 0 & 0 & 1 \\ -2/3 & -5/3 & -0.25 & 1 \end{bmatrix} \qquad B_2 = \begin{bmatrix} 0 & 2 \\ 0.25 & 0.8 \\ 0 & 0 \\ 1 & 1 \end{bmatrix}$$

$$C_2 = \begin{bmatrix} 1 & 0 & 0 & 0 \\ 0 & 0 & 1 & 0 \end{bmatrix}$$

Table 2: Parameter values for model 2

In the first model, the entries of the matrices have been chosen such that the third row of the Hankel matrix $H_{N,M}$ is nearly in the linear span of the first two rows. Hence the best structural vector of the process is $h_1 = (1,2,4,6)$.

The second model has been taken from a paper by H. El Sherief and N. K. Sinha [19]; using their own method, they find the structural vector $h_2 = (1,2,3,4)$ for this model.

In the sequel, the method of Ljung and Rissanen will be referred to as the LR method, the variant that we have introduced in section 4 will be called WG1, while the QR factorization procedure will be called method WG2.

Order	Structural vector	LR method Complexity	WG1 method trace	WG1 method J	WG2 method selected row	WG2 method |pivot|
3	(1,2,3)	4.37	4.27			
3	(1,2,4)	3.72 ←	4.03 ←	1.33	1	0.408
					2	0.054
4	(1,2,3,4)	7.49	21.52		4	0.110
4	(1,2,4,6)	5.00 ←	4.14 ←	1.04	6	0.104
4	(1,2,3,5)	6.72			8	0.100
					10	0.105
5	(1,2,3,4,6)	8.71	21.68			
5	(1,2,4,6,8)	7.17 ←	7.01 ←	1.40		
5	(1,2,3,4,5)	9.86				
5	(1,2,3,5,7)	9.08				

Table 3: Methods LR, WG1 and WG2 applied to model 1

Table 3 shows the results of the three methods applied to the first model. Note that in the LR method, the program computes the complexity of all admissible structures for each different order; method WG1, on the other hand, is recursive: for an order n, only those structural vectors are considered which contain all the rows selected in the optimal $(n-1)$-th order structural vector. The stopping criterion J used with WG1 has been defined at the end of section 4.

All three methods reject the odd rows in the structural vector. (Recall that the first p rows are always chosen following condition 2 of section 3). The LR method gives no estimate of the order. In the WG1 method, the criterion $J = \frac{1}{n}\,\mathrm{tr}(A^1)^{-1}$ is minimum for $n = 4$, while in method WG2, after the triangularization of row 4 (third step of the procedure) the decrease of the pivots is not significant anymore, which suggests an order 3.

Finally we add that the use of Akaike's canonical correlation method [3] on this model leads to the structural vector $h = (1,2,4)$.

Order	Structural vector	LR method Complexity	WG1 method trace	J	WG2 method selected row	|pivot|
3	(1,2,3)	0.92 ←	2.72			
3	(1,2,4)	1.14	2.28 ←	0.76	1	0.735
					2	0.964
4	(1,2,3,4)	2.06 ←	4.89 ←	1.22	4	0.230
4	(1,2,4,6)	2.68	9.22		3	0.063
4	(1,2,3,5)	4.08			5	0.062
					7	0.063
5	(1,2,3,4,5)	7.17	7266.8			
5	(1,2,3,4,6)	3.76 ←	15.11 ←	5.04		
5	(1,2,4,6,8)	7.91				
5	(1,2,3,5,7)	6.24				

Table 4: Methods LR, WG1 and WG2 applied to model 2

In table 4 we see that the three methods lead to a structural vector $h = (1,2,3,4)$ if the supposed order is 4, but if $n = 3$, the LR method leads to $h = (1,2,3)$ while the other two methods indicate $h = (1,2,4)$. The estimation of the order in WG1 leads to $n = 3$, while visual inspection of the pivots in WG2 leads to $n = 4$. Again, the simulation of Akaike's method leads to $h = (1,2,4)$.

The conclusion we have drawn from our simulations is that in most cases (with results of Table 4 being the sole exception) method WG1 seems to give the best estimate of the order of the model. We recall that Akaike's method gives an order estimate but does not select a "best basis," while the LR method does not estimate the order but selects a "best basis" within a prescribed order. As for method WG2, a better criterion than the visual inspection of the decrease of the pivots could probably be found.

We believe that all three methods give fairly good results for the determination of the structure, the advantage of the methods WG1 and WG2 being that they provide some estimate of the order as well, which avoids the fitting of too many parametrizations.

8. CONCLUSIONS. We have shown that the problem of specifying identifiable parametric structures for multivariable systems can be solved by a factorization of the probability map in such a way as to define a finite set of invariants which completely characterize the process. Proceeding in this way we have constructed a family of overlapping parametrizations which completely cover the set of finite-dimensional minimal-order systems. Since a given process can in general be represented by different overlapping parametrizations, the question then arises as to whether some parametrizations might yield more accurate parameter estimates than others. Our main result is that all overlapping parametrizations yield asymptotically the same value to the determinant of the information matrix. Therefore, when a prediction error identification method is used for the estimation of the parameters, all overlapping parametrizations will give the same value to the determinant of the asymptotic error covariance matrix. This is an asymptotic result. It does not imply that some structures might not be better than others when only a finite data record is available. Two heuristic selection schemes that can be used for the finite data case have been presented.

REFERENCES

1. D. G. Luenberger, *Canonical forms for linear multivariable systems,* IEEE Trans. Aut. Cont., vol. 12, pp. 290-293, 1967.

2. M. J. Denham, *Canonical forms for the identification of multivariable linear systems,* IEEE Trans. Aut. Cont., vol. 19, pp. 646-656, 1974.

3. H. Akaike, *Canonical Correlation Analysis of Time Series and the Use of an Information Criterion,* System Identification: Advances and Case Studies, (R. Mehra and D. Lainiotis, eds.), Academic Press, 1976.

4. J. Rissanen, *Basis of invariants and canonical forms for linear dynamic systems,* Automatica, vol. 10, pp. 175-182, 1974.

5. R. P. Guidorzi, *Invariants and canonical forms for systems structural and parametric identification*, Automatica, vol. 17, pp. 117-133, 1981.

6. K. Glover, J. C. Willems, *Parametrization of linear dynamical systems, canonical forms and identifiability,* IEEE Trans. Aut. Cont., vol. 19, pp. 640-646, 1974.

7. L. Ljung, J. Rissanen, *On canonical forms, parameter identifiability and the concept of complexity,* 4th IFAC Symp. on Identification and System Parameter Estimation, Tbilisi, USSR, 1976.

8. A. J. M. van Overbeek, L. Ljung, *On line structure selection for multivariable state space models,* 5th IFAC Symp. on Identification and System Parameter Estimation, Darmstadt, FRG, 1979.

9. V. Wertz, M. Gevers, E. Hannan, *The determination of optimum structures for the state space representation of multivariate stochastic processes,* submitted for publication.

10. J. Rissanen, *Estimations of structure by minimum description length,* Proc. Intern. Workshop on Rational Approximations for Systems, Leuven, Belgium, 1981.

11. M. Deistler, E. J. Hannan, *Some properties of the parametrizations of ARMA systems with unknown order,* to appear in J. Multivariate Analysis.

12. G. Picci, *Some numerical aspects of multivariable systems identification,* Proc. of the Workshop on Numerical Methods for Systems Engineering Problems, Lexington, Kentucky, June 1980.

13. B. D. O. Anderson, M. R. Gevers, *Overlapping state-space and ARMA parametrizations for multivariable systems,* in preparation.

14. M. van Emden, *Analysis of complexity,* Math. Cent. Tracts, 35, Amsterdam, 1971.

15. V. Wertz, *Détermination de la structure de processes multivariés,* Department of Electrical Engineering, Louvain University, IRSIA Report 79-80.

16. G. H. Golub, V. Klema, G. W. Stewart, *Rank degeneracy and least squares problems,* Tech. Reprt. Stan. CS76.559, Stanford University, August 1976.

17. G. H. Golub, G. P. H. Styan, *Numerical computations for univariate linear models,* J. Stat. Comp. Simul., 1973, vol. 2, pp. 253-274.

18. T. W. Anderson, *An introduction to multivariate statistical analysis,* Wiley, 1958.

19. M. El Sherief, N. K. Sinha, *Determination of the structure of a canonical model for the identification of linear multivariable systems,* IFAC Symp. on Identification and Parameter Estimation, Darmstadt, FRG, Sept. 1979.

LABORATOIRE D'AUTOMATIQUE ET D'ANALYSE DES SYSTÈMES
LOUVAIN UNIVERSITY, BÂTIMENT MAXWELL
B-1348 LOUVAIN-LA-NEUVE
BELGIUM

LIE THEORY OF TRANSFORMATION GROUPS AND THE PARAMETERIZATION AND IDENTIFICATION OF LINEAR SYSTEMS

Robert Hermann and Clyde F. Martin

1. INTRODUCTION. Since the Ames Conference on Geometric Control in 1976 [1], it has become widely recognized that the theory of systems is closely linked to algebraic and differential geometry and Lie group theory. The NATO Advanced Study Institute at Harvard University [16] reinforced this recognition and emphasized that the theory of linear systems, algebraic geometry and algebraic Lie groups is deeply linked. The 1981 APSM Workshop is designed to build upon this insight and move on to consider the cross-disciplinary area of system identification. These notes explore some of the possible applications of the Lie theory of transformation groups to system identification.

Let us briefly recall how Lie group theory enters into linear system theory [2]. Let

$$X,\ U,\ Y$$

be finite dimensional vector spaces over a scalar field. (The real of complex numbers will suffice.) A (linear, time-invariant) *input-output* system is (for simplicity) a system of differential equations (or sometimes difference equations):

$$\frac{dx}{dt} = Ax + Bu$$
$$y = Cx \quad , \tag{1.1}$$

where $A : X \to X$, $B : U \to X$, $C : X \to Y$ are linear maps. Let

$$Z = X \times U \times Y \times T$$

(T = time interval $0 \le t < \infty$). Each solution $t \to (x(t),u(t),y(t))$ of (1.1) defines a curve, i.e., a one-dimensional submanifold of Z:

$$t \to (x(t),u(t),y(t),t) \quad . \tag{1.2}$$

Let σ denote the collection of these curves. (Thus, σ is the system.) Let Σ be the set of all σ's. Σ may be *parameterized*

$$(A,B,C) \in L(X,X) \times L(U,X) \times L(X,Y) \quad .$$

One may then consider the group G of all invertible *linear* maps

$$Z \to Z$$

that map the curves constituting one system to the curves constituting another system. In this way, we obtain a transformation group acting on the space Σ. We can then pose standard questions of Lie transformation group theory [3,4], such as determining the orbits of G and its subgroups acting on Σ, determining the orbit spaces, etc. In fact, notice that this is just a special case of Lie's *general* conception [5] of how transformation groups act on systems of differential equations!

We begin this work with a survey of some standard concepts of the theory of groups of diffeomorphisms acting on manifolds.

2. GENERALITIES ABOUT ORBITS AND ORBIT SPACES. As we will see, many of the standard problems in system theory involve computing orbits and orbit spaces of Lie groups that act as transformation groups on manifolds. Let's consider a few general concepts, assuming that the reader knows the rudiments of Lie group theory and manifold differential geometry [6,15].

Let Z be a manifold and let G be a group. A *transformation group action* of G on Z is a mapping $G \times Z \to Z$, denoted by $(g,z) \to gz$, such that the following conditions are satisfied:

$$g_1(g_2 z) = (g_1 g_2)z$$

for $g_1, g_2 \in G$, $z \in Z$

$$1 \cdot z = z$$

for all $z \in Z$, where "1" denotes the identity element of G.

Suppose such a transformation group action is given. Introduce an equivalence relation on Z as follows:

$$(z_1 \sim z_2) \text{ iff there is a } g \in G \text{ such that } gz_1 = z_2 \quad .$$

The equivalence classes are then subsets of the form Gz, i.e., transforms under Gz of a point of Z. Each such subset is an *orbit* of G; if Gz is the *stability* or *isotropy subgroup* of G at z, i.e., the set of $g \in G$ such that $gz = z$, then the orbit is identified with the coset space G/G^z. In particular, each orbit is a *submanifold* of Z.

The *orbit space* of G acting on Z is the quotient space of this equivalence relation, denoted as $G\backslash Z$. $G\backslash Z$ is then a space, each of whose points is an orbit of the action of G. The quotient map $\pi : Z \to G\backslash Z$ is the map that assigns to each point $x \in Z$ the orbit Gx on which it lies. In general, $G\backslash Z$ is *not* a manifold. In many examples that are common in pure and applied mathematics, $G\backslash Z$ is a union of manifolds--something like a polyhedron, i.e., a bunch of manifolds glued together in some appropriate way. A typical example is when Z is the set of $n \times n$ nilpotent matrices and $G = G\ell(n)$. Then $G\backslash Z$ is not a manifold but is the union of manifolds $G\backslash X_\sigma$ where X_σ is the set of nilpotent matrices similar to the nilpotent matrix

$$\begin{pmatrix} S_{\sigma_1} & & & 0 \\ & \cdot & & \\ & & \cdot & \\ & & & \cdot \\ 0 & & & S_{\sigma_i} \end{pmatrix}$$

S_{σ_i} is the $\sigma_i \times \sigma_i$ shift matrix and $\sigma_1 + \ldots + \sigma_i = n$. In this case $G\backslash X_\sigma$ is a single point and hence $G\backslash Z$ is a finite but the topology is not Hausdorf. See [17] for a discussion of this phenomena.

However, let us consider one case where $G\backslash Z$ is a manifold.

DEFINITION. *The action of* G *on* Z *is said to be* regular *if* $G\backslash Z$ *can be given a manifold structure such that the quotient map*

$$\pi : Z \to G\backslash Z$$

is a submersion, *i.e., such that the linear map*

$$\pi_* : T(Z) \to T(G\backslash Z)$$

induced by π *on tangent vectors is* onto.

DEFINITION. *An orbit* Gz *of* G *acting on* Z *is* regular *if it is contained in an open subset* U *of* Z *such that:*

U *is invariant under the action of* G.

G *acts in a regular way on* U.

REMARK. By definition, the set of regular points is an open subset of Z that is invariant under the action of G. In certain cases (e.g., the case where G leaves invariant a *complete* positive Riemannian metric [15], in which case the orbit structure looks much as it does in the classic case of the action of a compact group [3], and certain types of complex analytic and/or algebraic actions), one can prove that the regular points are *dense*, and their compliment is a set of lower dimension. The algebraic case is of particular importance in linear system theory.

3. ORBIT TYPES. Suppose that the group G acts on a space Z, as in Section 2. For $z \in Z$, G^z denotes the stability subgroup of G at z.

DEFINITION. *Points* z *and* z' *of* Z *have the* same orbit type *if there is a* $g \in G$ *such that*

$$G^z = gG^{z'}g^{-1} \quad , \tag{3.1}$$

i.e., if the stability subgroups at the two points are conjugate *as subgroups of* G.

THEOREM 3.1. *The relation*

$z \sim z'$ *iff* z *and* z' *have the same orbit type*

is an equivalence relation.

The proof of Theorem 3.1 is easy and is left to the reader. This result shows that there is a partitioning of Z into subsets--the equivalence classes, i.e., the points with the same orbit class. This structure plays a basic role in "symmetry breaking" which occurs in many areas of applied mathematics and "bifurcation theory" [13].

4. A TRANSFORMATION GROUP ACTION DETERMINES AN INTERTWINING MAPPING OF X INTO THE SPACE OF ALL SUBGROUPS OF G. Let $S(G)$ denote the space, each of whose "points" is a *subgroup* of G. The mapping mentioned in the title of this section is then the map ϕ defined by

$$z \to G^z \in S(G) \quad . \tag{4.1}$$

Let G act on $S(G)$ by conjugation:

$$(g,S) \to gSg^{-1} \quad \text{if } g \in G \text{ and } S \text{ is a subgroup of } G \quad . \tag{4.2}$$

THEOREM 4.1. *The mapping* $\phi : Z \to S(G)$ *indicated in* (4.1) *is an* intertwining map *for the action of* G.

<u>Proof</u>. The "intertwining" property means, abstractly, that

$$g\phi(z) = \phi(gz) \quad \text{for } g \in G, \ z \in Z \quad . \tag{4.3}$$

If $\phi(z) = G^z$, then, using (4.2)

$$g\phi(z) = gG^zg^{-1} \quad ,$$

which is also equal to G^{gz}, which is $\phi(gz)$, i.e., (4.3) is indeed satisfied. Q.E.D.

We can now interpret the "orbit type" concept from this more general point of view. The action of G on S(G) determines an equivalence relation on S(G), the orbits. One can now "pull back" this equivalence realtion via ϕ, i.e.,

$$z \sim z' \text{ iff } \phi(z) \sim \phi(z') \quad .$$

This abstract construction, applied to this case, gives the partitioning of Z by orbit type.

Of course this map

$$\phi : Z \to S(G)$$

is important for many purposes in transformation group theory. It is very important for many applications (in both pure and applied work) to be able to compute it as explicitly as possible. This is often a very difficult algebraic problem. Often, in Lie group situations, one can study S(G) and the action of G on it in the context of the theory of *deformations* of Lie groups and algebras [7-12].

So far, we have not supposed any particular structure on Z and G. Of course, the problem at hand will usually suggest natural hypotheses about the structure. For example:

Z and G are finite.

Z and G are discrete.

Z and G are topological spaces, the maps $G \times G \to G$, $G \times Z \to Z$ being continuous.

Z and G are manifolds, the maps $G \times G \to G$, $G \times Z \to Z$ being differentiable.

Z and G are complex manifolds, the maps $G \times G \to G$, $G \times Z \to Z$ being complex analytic maps, and so on.

In the "Lie" situation, we can further define

$$Z \to \mathfrak{G}^z \in \text{space of linear subspaces of } \mathfrak{G},$$

where $\mathfrak{G}$ is the Lie algebra of G (a finite dimensional vector space), and $\mathfrak{G}^z$ is the Lie algebra of the Lie subgroup of G^z. This is usually a more amenable object. These concepts also appear in applied mathematics in terms of "bifurcations" and "catastrophes" [13].

5. THE ORDER RELATION DEFINED BY THE CLOSURE OF THE ORBIT TYPES IN THE FINITE DIMENSIONAL LIE SITUATION. Suppose now that Z is a (finite dimensional) manifold, that G is a Lie group, and that the action $G \times Z \to Z$ of G on Z is differentiable. For $z \in Z$, let

$O(z)$ = set of all points z' of Z which have the same orbit type as z.

We can then consider its *closure* in Z, in the sense of pointset topology--the topology on Z is that given by the manifold. Denote this closure by

$$\overline{O(z)} \quad .$$

DEFINITION. *Given* $z,z' \in Z$, *let us say that*

$$\boxed{\begin{array}{c} O(z') \le O(z) \ \text{iff} \\ z \in \overline{O(z')} \end{array}} \tag{5.1}$$

REMARK. At this level of generality, it is not true that $O(z) \le O(z')$ is a genuine order relation. The transitivity condition

$$(z \le z') \text{ and } (z' \le z'') \Rightarrow z \le z''$$

seems to be nontrivial.

Here are some examples.

EXAMPLE 1. $O(n,R)$ acting on the unit sphere in R^{n+1}.

Let R^{n+1} denote the (n+1)-dimensional real Euclidean space, with the usual scalar dot product. Let S^n be the set of all $z \in R^{n+1}$ such that

$$||z||^2 \equiv z \cdot z = 1 \quad .$$

$O(n+1,R)$ is the group of linear orthogonal transformations $g : R^{n+1} \to R^{n+1}$ such that

$$||gz|| = ||z||$$

$$\text{for } z \in R^{n+1} \quad .$$

$O(n+1,R)$ acts transitively on S^n. The stability subgroup at one point is $O(n,R)$. Explicitly, let

$$z^0 = (0,0,\ldots,1) \quad .$$

We will call this the north pole. (Think of our R^3 as the set of all ordered (z_1,z_2,z_3). The plane $z_3 = 0$ is the *equator* of the earth. Then, $(0,0,1)$ is indeed the "North Pole.")

$O(n,R)$ = set of all $g \in O(n+1,R)$ such that

$$gz^0 = z^0 \quad .$$

Let $G = O(n,R)$, acting on S^n. For $z \in S^n$, G^z is the group of all orthogonal transformations $g : R^{n+1} \to R^{n+1}$ such that:

$$g(z^0) = z^0$$

and

$$g(z) = z \quad .$$

Now, if z and z^0 are linearly independent, then g is determined by its action on the orthogonal complement in R^{n+1}, i.e., g belongs to $O(n-1,R)$. Hence, if

$$z \neq \pm z^0 \quad ,$$

then G^z is conjugate (within G) to $O(n-1,R)$. But, of course, when z becomes $\pm z^0$, G^z "jumps" to become G.

Thus, there are *two* orbit types for this action; those for which the stability group is $O(n-1,R)$ or $O(n,R)$. Of course, the latter lie in the *closure* of the former, i.e., the space of orbit types consists of two elements e_0, e_1 with

$$e_0 \geq e_1 \quad .$$

EXAMPLE 2. $Gl(n,R)$ acting on R^n.

z denotes an element of R^n. g is an $n \times n$ real matrix of non-zero determinant. There are then, again, two orbit types, with the stability group $GL(n,R)$ itself and the identity subgroup of this group. The former is in the closure of the latter.

EXAMPLE 3. $G = O(n,R)$ acting on the space Z of $n \times n$ real symmetric matrices.

The action is

$$(g,z) = gzg^{-1} = gzg^T \quad .$$

Let D be the subset of Z consisting of the *real diagonal* matrices. Then,

$$GD = Z \quad .$$

(This is the classical "Gram-Schmidt" theorem that any symmetric matrix is orthogonally similar to a diagonal one.) Now, it also is easy to prove that two diagonal matrices

$$\delta = \begin{pmatrix} \lambda_1 & & & \\ & \cdot & & \\ & & \cdot & \\ & & & \cdot \\ & & & \lambda_n \end{pmatrix}$$

$$\delta = \begin{pmatrix} \lambda_1' & & & \\ & \cdot & & \\ & & \cdot & \\ & & & \cdot \\ & & & \lambda_n' \end{pmatrix}$$

are orthogonally conjugate if and only if their diagonal elements differ by a permutation.

Thus, the orbit space

$$G\backslash Z$$

is the orbit space of the action of the permutation group S_n on R^n.

Now, if

$$\delta = \begin{pmatrix} \lambda_1 & & & \\ & \cdot & & \\ & & \cdot & \\ & & & \cdot \\ & & & \lambda_n \end{pmatrix}$$

with diagonal entries all unequal, one sees that G^δ --the subgroup of $g \in O(n,R)$ such that

$$g\delta g^{-1} = \delta$$

consists of $g = \pm 1$, i.e.,

$G^\delta = C_2$, the cyclic group on two elements.

Let $P(n)$ be the partitions for the integer n:

$\pi \in P(n)$ is then a sequence $(\pi_1,\ldots,\pi_m)$ of positive integers such that

$$n = \pi_1 + \ldots + \pi_m \quad .$$

Let $D(\pi)$ be the subset of $\delta \in D$ of the form:

$$\delta = \begin{pmatrix} \left.\begin{matrix} \lambda_1 & & & \\ & \cdot & & \\ & & \cdot & \\ & & & \cdot \\ & & & & \lambda_1 \end{matrix}\right] \pi_1 \text{ entries} & & \\ & \left.\begin{matrix} \lambda_2 & & & \\ & \cdot & & \\ & & \cdot & \\ & & & \cdot \\ & & & & \lambda_2 \end{matrix}\right] \pi_2 \text{ entries} & \\ & & \begin{matrix} \cdot & & \\ & \cdot & \\ & & \cdot \end{matrix} \end{pmatrix}$$

In words, the diagonal elements of δ are equal according to the partition, *but that diagonal* elements in different "boxes" are unequal. It is then easy to see that:

$$G^{\delta} = O(\pi(1),R) \times \ldots \times O(\pi(m),R) \equiv G^{\pi} \quad .$$

This determines the order types: *they are one-one correspondence with the partitions* $P(n)$.

Let $\delta \in D(\pi)$. The closure $\overline{D(\pi)}$ is then determined by letting some of the unequal diagonal elements in different boxes become equal. This defines an order on $P(n)$:

If $\quad \pi = (\pi_1, \ldots \quad)$

and $\quad \pi' = (\pi'_1, \ldots \quad)$,

then $\quad \pi' \leq \pi \quad$ iff

$G(\pi') \subset G(\pi) \quad .$

In this case, we see that the ordering on orbit types *agrees* with a more natural ordering--inclusion among subgroups of $O(n,R)$. In fact, those orderings are of great interest from the point of view of that branch of mathematics called "combinatorics." There is another ordering on partitions, the so-called dominance ordering that occurs frequently. Let π and γ be two partitions of n in decreasing order. Then $\pi \geq \gamma$ iff

$$\begin{cases} \pi_1 \geq \gamma_1 \\ \pi_1 + \pi_2 \geq \gamma_1 + \gamma_2 \\ \quad \vdots \\ \pi_1 + \ldots \pi_n \geq \gamma_1 + \ldots + \gamma_n \quad . \end{cases}$$

This ordering occurs in orbit closures connected with conjugation in nilpotent matrices, the feedback group acting in systems and in general is related on the Bruhat decomposition of an algebraic group [17].

In some of these examples, G is a Lie group acting on a manifold X such that there is a *complete*, positive Riemannian metric on X which is invariant under G. We can then use *Riemannian geometric* concepts to study the "orbit type" concept. A certain amount of this has been done in *Differential Geometry and the Calculus of Variations* [15]. Let us now review some of this material and try to push it further.

6. ORBITS AND ORBIT TYPES FOR CLOSED CONNECTED GROUPS OF ISOMETRIES OF COMPLETE POSITIVE RIEMANNIAN MANIFOLDS. PRINCIPAL ORBITS. Let X be a complete, positive definite Riemannian manifold. Let G be a group of isometries of the metric on X. Suppose that G is *connected* and *closed* in the group of all isometries of the metric.

Each orbit

$$Gx$$

of G is then a *closed, regularly embedded* submanifold of X. It then has a *geodesic tubular* neighborhood

$$Gx \subset U \subset X$$

with the following properties:

a) Each point $x' \in U$ can be joined to Gx by a *unique* geodesic of *minimal length.*

b) $\overline{U} = X$, i.e., every point in X is in the closure of U. In fact, if x' is an arbitrary point of X and if

$$t \to \sigma(t) \quad , \quad 0 \le t \le 1$$

is a geodesic curve of minimal length joining x' to Gx ;

$$\sigma(0) = x'; \quad \sigma(1) \in Gx ;$$

then $\sigma(t) \in U$ for $0 < t \le 1$. Thus, for x' in U, such a tubular neighborhood of Gx

$$G^{x'} \subset G^{x} \quad . \tag{6.1}$$

<u>Proof of 6.1</u>. Let $g \in G^{x'}$,

$$g \in G \text{ and } gx' = x' \quad .$$

If g were *not* in G^{x}, i.e., $gx \neq x$, σ and $g\sigma$ would be two geodesics of equal length, both joining x' to Gx, which would contradict the "tubular *geodesic* neighborhood" property of U.

This property suggests some useful definitions.

DEFINITION. *A point* x^0 *of* X *is of* principal orbit type *for the action of* G *if the following condition is satisfied:*

> G^{x^0} *has a minimal dimension and a minimal number of components, as compared to* G^x, *as* x *varies over all of* X.

Let us suppose now that Gx^0 is such a principal orbit and that U is a geodesic tubular neighborhood.

THEOREM 6.1. *For* $x \in U$, Gx *is also a principal orbit.* G^x *is conjugate to* G^{x^0} *within* G. *In other words, all elements in* U *have the same orbit type.*

Proof. Given $x \in U$, it can be joined to Gx^0 by a geodesic σ of minimal length, which lies completed in U. By at most replacing x^0 with another point on the same orbit, we may suppose that σ runs from x to x^0, i.e.,

$$\sigma(0) = x \; ; \quad \sigma(1) = x^0 \quad .$$

For $0 \leq t \leq 1$, we have, as we have seen,

$$G^{\sigma(t)} \subset G^{x^0} \quad .$$

However, by the *minimality* property we used to define x^0, as a point on a principal orbit, we must have:

$$G^{\sigma(t)} = G^{x^0} \quad \text{for} \quad 0 \leq t \leq 1 \quad .$$

Q.E.D.

7. INTERTWINING MAPS FOR TRANSFORMATION GROUP ACTIONS. Let G and G' be groups acting as transformation groups on two manifolds X and X'. A pair of maps $\phi : X \to X'$, $\alpha : G \to G'$ are said to be *intertwining maps* (relative to the action of G) if the following condition is satisfied:

$$\phi(gx) = \alpha(g)\phi(x) \qquad \text{for } g \in G, \; x \in X \quad . \tag{7.1}$$

α is a surjective *homomorphism* of groups, i.e.,

$$\alpha(g_1 g_2) = \alpha(g_1)\alpha(g_2) \tag{7.2}$$

and is onto.

Let G and G' be groups that act as transformation groups on manifolds X and X'. Let $\phi : X \to X'$, $\alpha : G \to G'$ be maps that intertwine the action of G. Our aim in this section is to study certain simple consequences on the orbits of G of the existence of the intertwining map.

Let $G\backslash X$ and $G'\backslash X'$ denote the orbit spaces, i.e., a "point" of $G\backslash X$ is a subset of X that is an orbit of G. Let $\pi : X \to G\backslash X$ and $\pi' : X' \to G'\backslash X'$ be the quotient maps, which assign to each point of X and X' the orbits of G on which they lie.

THEOREM 7.1. *ϕ maps an orbit of G on X into an orbit of G' on X'.*

Proof. For $x \in X$, using the intertwining property (7.1), we have:

$$\phi(Gx) = G'\phi(x) \quad ,$$

which shows that, for each orbit Y of G on X, $\phi(Y)$ is an orbit of G' on X'.

When there is little likelihood of confusion, we will also denote by ϕ, the quotient map, induced on the orbit space: $G\backslash W \to G'\backslash X'$. Thus, we have a "commutative diagram" of maps:

$$\begin{array}{ccc} X & \xrightarrow{\phi} & X' \\ {\scriptstyle\pi}\downarrow & & \downarrow{\scriptstyle\pi'} \\ G\backslash X & \xrightarrow{\phi} & G'\backslash X' \end{array}$$

If $Y' \subset X'$ is an orbit of G', $\phi^{-1}(Y')$ is invariant under the action of G, since, for $g \in G$,

$$\begin{aligned} \phi(g\phi^{-1}(Y')) &= \alpha(g)\phi\phi^{-1}(Y') \\ &= \alpha(g)Y' \\ &= Y' \quad . \end{aligned}$$

If $\phi^{-1}(Y')$ is non-empty, let Y be an orbit of G contained in $\phi^{-1}(Y')$. Then, $\phi(Y)$ is an orbit of G' that is contained in Y', hence must be equal to Y', since Y' is itself an orbit of G. In particular, we have proved:

THEOREM 7.2. *If $\phi(X) = X'$, then*

$$\phi(G\backslash X) = G'\backslash X' \quad .$$

Let us now study the relation between G-orbits $Y \subset X$ and $Y' \subset X'$ such that

$$\phi(Y) = Y' \quad .$$

Let y and y' be points of Y and Y' such that:

$$\phi(y) = y' \quad .$$

Let G^y and $G'^{y'}$ be the isotropy subgroups of G at y and y'. Y and Y' can then be identified with the coset spaces

$$G\backslash G^y \text{ and } G'\backslash G'^{y'} \quad .$$

THEOREM 7.3. $\alpha(G^y)$ *is a subgroup of* $G^{y'}$.

Proof. If $g \in G^y$, then $gy = y$. Hence,

$$\begin{aligned} \phi(gy) &= y' \\ &= \alpha(g)\phi(y) \\ &= y' \quad , \end{aligned}$$

i.e.,

$$\alpha(g) \in G'^{y'} \quad ,$$

hence,

$$\alpha(G^y) \subseteq G'^{y'} \tag{7.3}$$

Thus, we have proved:

THEOREM 7.4. *The map* ϕ *induces:* $Y \to Y'$ *may be identified with the mapping on cosets:*

$$G/G^y \to G'/G'^{y'} \quad , \tag{7.4}$$

with fiber (at the identity cosets) $G^{y'}/(G^y)$, *which is induced by the inclusion:* $\alpha(G^y) \subset G'^{y'}$.

REMARK. More precisely, the map (7.4) is defined as follows: An "element" of G/G^y is a subset of G of the form: gG^y, where g is an element of G. Such a subset is contained (because of (7.3) in the subset $\alpha(g)G'^{y'}$, i.e., an element of the coset space $G'/G'^{y'}$. This assignment

$$gG^y \to \alpha(g)G'^{y'}$$

(which is readily seen to be independent of the choice of the element g defining the coset) defines the map (7.4). The fiber of the map (7.4) above the identity coset consists of the gG^y such that $\alpha(g) \in G'^{y'}$. This is then just identified with the coset space $G'^{y'}/\alpha(G^y)$.

Now, let us study the case where

$$G' \text{ acts transitively on } Y' \quad . \tag{7.5}$$

Let y'_0 be a point of Y' such that:

$$Y' \text{ is identified with the coset space } G'/G'^{y'_0}$$

Set

$$Y_0 = \phi^{-1}(y'_0)$$

then

$$\alpha^{-1}(G'^{y'_0})Y_0 = Y_0 \tag{7.6}$$

and

$$GY_0 = Y \quad .$$

<u>Proof</u> of Eq. (7.7). If $y \in Y$, then there is a $g \in G$ (because of (7.5) such that

$$\alpha(g)\phi(y) = y'_0 \quad ,$$

i.e., $\phi(gy) = y'_0$ or $gy \in Y_0$.

Let $\pi_0 : Y_0 \to G\backslash Y$ be the map resulting from restricting the map π to Y_0. By (7.7), π_0 is onto $G\backslash Y$. Let us compute the fibers of π_0.

Suppose then that $y_1, y_2 \in Y_0$, with $gy_1 = y_2$. Then

$$\phi(gy_1) = \phi(y_2) \quad ,$$

or

$$\alpha(g)y'_0 = y'_0$$

or

$$\alpha(g) \in G'^{y'_0} \quad .$$

This proves the following result.

THEOREM 7.5. *If* (7.5) *is satisfied, and if* Y_0 *is the fiber of* ϕ *above a point* y'_0 *of* Y' *, then* π_0 *identifies the orbit space*

$$\alpha^{-1}(G'^{y'_0})Y_0 \text{ with } G\backslash Y \quad .$$

EXAMPLE. Suppose that G acts transitively on spaces Y_1 and Y_2. Set:

$$Y = Y_1 \times Y_2$$
$$Y' = Y_2 \quad . \tag{7.8}$$

The action of G on Y is the "diagonal" action, i.e.

$$g(y_1,y_2) = (gy_1,gy_2)$$

for $y_1 \in Y_1$, $y_2 \in Y_2$, $g \in G$. α is the identity map. Let $\phi : Y \to Y'$ be the Cartesian projection, i.e.,

$$\phi(y_1,y_2) = y_2 \quad .$$

Then, $G\backslash Y'$ consists of a single point, i.e., (7.5) is satisfied. Let y_0' be a point of Y_2. Then, Y_0 is the subset

$$Y_1 \times (y_0') \quad .$$

The action of $G^{y_0'}$ on Y_0 is then just the action of $G^{y_0'}$ on Y_1. If Y_1 is the coset space

$$G/G^{y_1}$$

then Theorem (7.5) shows that $G\backslash Y$ is identified with the orbits of $G^{y_0'}$ on G/G^{y_1}, which are just the space of double cosets

$$G^{y_0'} \backslash G/G^{y_1} \quad . \tag{7.9}$$

REMARK. An element of the "double coset space" (7.9) is a subset of G of the form:

$$G^{y_0'} g G^{y_1} \quad ,$$

with g an element of G. In other words, the double coset space (7.9) is the orbit space of G under the action $G^{y_0'} \times G^{y_1}$ on G defined by the following formula:

$$(g_1,g_2)(g) = g_1 g g_2^{-1} \quad , \tag{7.10}$$

where $g_1 \in G^{y_0'}$, $g_2 \in G^{y_1}$.

This identification of the orbit space of G on $Y_1 \times Y_2$ with a double coset space is one basic element in the theory of "induced representations" of groups. It also arises, as explained in [14], Chapter 12, in the decomposition into irreducible representations of vector bundle representations of Lie groups.

8. PRIMITIVE TRANSFORMATION GROUP ACTIONS. Let G be a Lie group, which acts as a transformation group on a manifold X.

DEFINITION. *The action of* G *on* X *is said to be* decomposible *is there is a manifold* X' *on which* G *acts, with*

$$0 < \dim X' < \dim X \quad ,$$

and an intertwining map $\phi : X \to X'$ *such that:*

$$\phi(X) = X' \quad .$$

Thus, the orbit spaces of decomposible mappings can be computed by computing the orbits on X of the isotropy subgroup of the action of G on X'. Those spaces X which are *not* decomposible then play a certain "primitive" role in the structure of all group actions. This motivates the following definition, which goes back to Sophus Lie's original work in the 19th century on transformation groups.

DEFINITION. *The action of* G *on a manifold* X *is said to be* primitive *if it is not decomposible, in the sense defined above.*

In order to get a more precise feeling for the meaning of "primitivity," suppose that we begin with the simplest case, namely that there G acts transitively on X. Suppose that the action of G is not primitive, i.e., there is an onto intertwining map

$$\phi : X \to X' \quad ,$$

with $0 < \dim X' < \dim X$. Then, the Theorem 7.2, G acts transitively on X'. Let $x' \in X'$, $x \in \phi^{-1}(x')$. Then,

$$G^x \subset G^{x'} \subset G \quad , \tag{8.1}$$

$$\dim G^x < \dim G^{x'} < \dim G \quad . \tag{8.2}$$

Conversely, if there is a *closed* subgroup $G^{x'}$ of G satisfying (8.2), then X' can be chosen as $G/G^{x'}$, thus providing a decomposition of X. This proves the following result:

THEOREM 8.1. *If* X *is the coset space* G/K, *with* K *a closed subgroup of* G, *then the action of* G *on* X *is primitive if and only if there is no closed subgroup* L *of* G *such that:*

$$K \subset L \subset G \tag{8.3}$$

$$\dim \mathfrak{K} < \dim \mathfrak{L} < \dim \mathfrak{G} \quad . \tag{8.4}$$

COROLLARY TO THEOREM 8.1. *If* G *is a connected group and* $\mathfrak{K}$ *is a maximal subalgebra of* $\mathfrak{G}$, *then* G *acts primitively on* G/K.

REMARK. Obviously, the property that G acts primitively on G/K is "almost" equivalent to the property that $\mathfrak{K}$ is a maximal subalgebra of $\mathfrak{G}$. The difference is basically a "global" question. In fact, in the 19th century, an ill-defined "local" notion of group action and "primitivity" was used that resulted in the equivalence of these two properties. What was not realized then was that there could be subalgebras $\mathfrak{L}$ of $\mathfrak{G}$ such that: $\dim \mathfrak{L} < \dim \mathfrak{G}$, but that the closure in G of the subgroup L of G generated by $\mathfrak{L}$ is all of G. In fact, very little is known even today about the precise circumstances under which this may happen. For example, this cannot happen if $\mathfrak{G}$ is a simple Lie algebra.

An example of imprimitive action in system theory arises in the study of the so-called "Riccati group" that acts on the parameters of the linear quadratic problem-- (A,B,Q,R,S) where (A,B) is a controllable pair, Q,R and S are positive matrices from the integral cost criteria

$$J(u) = \int_0^T x'Qx + u'Rudt + x(T)'Sx(T) \quad .$$

Garcia [18] has studied this action via the embedding of quintuple into the symplectic Lie algebra and viewing the Riccati group as a subgroup of the symplectic group.

9. ACTION OF LIE GROUPS ON THE INPUT SYSTEMS. After these generalities, let us return to the case of *linear systems*. We change notation to be in concordance with that in common use in linear system theory. For simplicity, we consider first only the systems of the following form:

$$\frac{dx}{dt} = Ax + Bu \tag{9.1}$$

$x \in X$, the *state space*

$u \in U$, the *input space*

$A \in L(X,X)$, $B \in L(U,X)$.

Let Σ be the space of systems of the form (9.1), i.e.,

$$\begin{aligned} \Sigma &= L(X,X) \times L(U,X) \\ &= \{(A,B)\} \quad . \end{aligned} \tag{9.2}$$

Let G be the group of invertible linear maps

$$X \oplus U \to X \oplus U \quad ,$$

which preserves the space of solutions of (9.1). Thus $g \in G$ is of the following partitioned form [15].

$$g \begin{pmatrix} x \\ u \end{pmatrix} = \begin{pmatrix} g_{11} & g_{12} \\ g_{21} & g_{22} \end{pmatrix} \begin{pmatrix} x \\ u \end{pmatrix}$$

$$= \begin{pmatrix} g_{11}x + g_{12}u \\ g_{21}x + g_{22}u \end{pmatrix}$$

thus, if $(x(t),u(t))$ is a solution of (9.1)

$$\begin{pmatrix} x'(t) \\ u'(t) \end{pmatrix} = g \begin{pmatrix} x(t) \\ u(t) \end{pmatrix}$$

is to be a solution of a similar equation with a *new* A',B'. Let us work out the conditions for this:

$$\frac{dx'}{dt} = g_{11} \frac{dx}{dt} + g_{12} \frac{du}{dt}$$

$$= g_{11} (Ax + Bu) + g_{12} \frac{du}{dt}$$

if x' is to solve a similar equation parameterized by (A',B')

$$A'x' + B'u' = A'(g_{11}x + g_{12}u) + B'(g_{21}x + g_{22}u) \quad .$$

Thus,

$$g_{12} = 0 \tag{9.3}$$

$$g_{11}A = A'g_{11} + B'g_{21} \tag{9.4}$$

$$g_{11}B = B'g_{22} \tag{9.5}$$

THEOREM 9.1. *Let* G *be the group of invertible linear transformations of* $X \oplus U$ *of the form:*

$$g\begin{pmatrix} x \\ u \end{pmatrix} = \begin{pmatrix} g_{11} & 0 \\ g_{21} & g_{22} \end{pmatrix}\begin{pmatrix} x \\ u \end{pmatrix} \tag{9.6}$$

Then, g *maps* Σ *into itself via the following formula*

$$g(A,B) = (g_{11}Ag_{11}^{-1}, g_{11}Bg_{21}, g_{11}Bg_{22}^{-1}) \quad . \tag{9.7}$$

One can similarly calculate the action of the Lie algebra $\mathfrak{G}$ of G on Σ. First, $\mathfrak{G}$ is identified with the vector space of partitioned linear maps: $\alpha : X \times U \to X \times U$ of the form

$$\alpha = \begin{pmatrix} \alpha_{11} & 0 \\ \alpha_{21} & \alpha_{22} \end{pmatrix} \tag{9.8}$$

α then acts as a vector field V_α on $L(X,X) \times L(U,X)$, i.e., as a map

$$V_\alpha : L(X,X) \times L(U,X) \to L(X,X) \times L(U,X) \quad .$$

The explicit formula for V_α is readily obtained from (8.7):

$$V_\alpha(A,B) = ([\alpha_{11},A] - \alpha_{11}B + B\alpha_{21}, \alpha_{11}B - B\alpha_{22}) \quad . \tag{9.9}$$

The explicit calculation of the isotropy groups is quite involved and has never been alone explicitly. However using indirect methods, much can be said about the orbit structure and the orbit space. In particular, Brockett [19], determines the dimensions of the orbits (and hence the dimensions of the isotropy groups). Byrnes [20] uses a very clever argument to obtain the same results. Krishnaprasad and Martin [21] obtain the most detailed results on the orbit structure by using the concepts of flag manifolds and an associated canonical form. Hazewinkel and Martin [17] completely determine the ordering on the orbit closures and show that it is related in a very deep way with a variety of other phenomena, such as the similarity equivalence on nilpotent matrices (nilpotent varieties), the representations of the symmetric group (Weyl groups), Schubert calculus and Bruhat decomposition of the $G\ell(N,\mathbb{R})$.

The feedback group's action is deceptively easy to analyze since there are only finitely many orbits. The ordering relation of Section 5 is then relatively easy to analyze. On the other hand, if we consider the following action on *controllable pairs* (A,B) the situation is quite different.

$$(P(A,B)) \to (PAP^{-1}, PB) \quad .$$

Let M denote the controllable systems of fixed input and state dimensions. An easy calculation shows that the isotropy group is constant and is the identity for all controllable pairs (A,B). For suppose

$$PAP^{-1} = A$$
$$PB = B$$

then if $R(A,B) = [B,AB,\ldots,A^{n-1}B]$ then $PR(A,B) = R[PAP^{-1},PB] = R(A,B)$. Since $R(A,B)$ is of full rank P must be the identity. In this case there is exactly one orbit type and it follows that each orbit is closed. If we consider L, the space of all pairs (A,B) of fixed dimension then, since M is described by a set of polynomial inequalities, M is open and dense in L. However, the relation of the orbits corresponding to non-controllable systems is very complex and is at best incompletely understood. Khadr [22] has begun a systematic study of these relations, but there is a reason to believe that a complete description is not possible.

Of interest in identification is the concept of minimal parameterizations of systems. This is really a question of the structure of the orbit space of M modulo the $G\ell(n)$ action. The structure is not completely known, but has been studied in some aspects by Hazewinkel [23] and Byrnes and Hurt [24]. Their main thrust was to show the non-existence of continuous canonical forms except in the single input case. Both papers ultimately showed that while $M/G\ell(n)$ is a manifold (projective variety) that it is not equivalent to an Euclidean space. However, a complete description of the topology has yet to be given.

REFERENCES

1. R. Hermann and C. Martin, *Proceedings of Ames (NASA) 1976 Conference on Geometric Control*, Math Sci Press, Brookline, MA, 1977.

2. R. Hermann, *Linear Systems Theory and Introductory Algebraic Geometry*, Vol. VIII, Interdisciplinary Mathematics, Math Sci Press, Brookline, MA, 1974.

3. D. Montgomery and L. Zippen, *Topological Transformation Groups*, Wiley, New York, 1955.

4. *Sophus Lie's 1880 Transformation Group Paper*, comments and additional material by R. Hermann, Lie Groups: History Frontiers and Applications, Vol. 1, Math Sci Press, Brookline, MA, 1975.

5. *Sophus Lie's 1884 Differential Invariants Paper*, comments and additional material by R. Hermann, Lie Groups: History, Frontiers and Applications, Vol. 3, Math Sci Press, Brookline, MA, 1975.

6. W. Boothby, *An Introduction to Differentiable Manifolds and Riemannian Geometry*, Academic Press, 1975.

7. R. Richardson, Acta Math., circa 1973.

8. R. Hermann, *Analytic continuation of group representations*, Comm. Math. Phys., Vol. 2, pp. 251-270 (1966).

9. R. Hermann, *Analytic continuation of group representations, II*, Comm. Math. Phys., Vol. 3, pp. 53-74 (1966).

10. R. Hermann, *Analytic continuation of group representations, III*, Comm. Math. Phys., Vol. 3, pp. 75-97 (1966).

11. R. Hermann, *Analytic continuation of group representations,* Comm. Math. Phys., Part IV, Vol. 5, pp. 131-156 (1967); Part V, Vol. 5, pp. 157-190 (1967); Part VI, Vol. 6, pp. 205-225 (1967).

12. R. Hermann, *Physical Aspects of Lie Group Theory,* Univ. of Montreal Press, Montreal, 1974.

13. D. Sattinger, *Topics in Stability and Bifurcation Theory*, Springer Lecture Notes #309, 1973.

14. R. Hermann, *Lie Groups for Physicists,* W. A. Benjamin, 1966.

15. R. Hermann, *Differential Geometry and the Calculus of Variations,* second edition, Math Sci Press, Brookline, MA, 1976.

16. C. Byrnes and C. Martin, eds., *Geometrical Methods for the Theory of Linear Systems,* Reidel, Dordrecht, 1981.

17. M. Hazewinkel and C. Martin, *Representations of the symmetric group, the specialization order, systems and Grassman manifolds,* preprint.

18. M. Garcia, *Linear Hamiltonian Systems and Transformation Groups in Linear Quadratic Optimal Control: A Lie Theoretic Approach,* Ph.D. dissertation, Case Western Reserve University, Cleveland, Ohio, 1982.

19. R. Brockett, *The Geometry of the Set of Controllable Linear Systems,* Research Report of Automatic Control Laboratory, Faculty of Engineering, Nagoya University, Vol. 24, pp. 1-7 (June 1977).

20. C. Byrnes, *On the control of certain infinite-dimensional deterministic systems by algebra-geometric techniques*, Amer. J. of Math., Vol. 100, pp. 1333-38 (1978).

21. P. Krishnaprasad and C. Martin, *Families of systems and deformations,* preprint.

22. A. Khadr, *On the Limiting Behavior of Families of Linear Systems: An Algebraic Geometric Approach,* Ph.D. dissertation, Case Western Reserve University, Cleveland, Ohio, 1981.

23. M. Hazewinkel, *Moduli and canonical forms for linear dynamical systems, II: The topological case*, Math. Syst. Theory, Vol. 10, pp. 363-385 (1977).

24. C. Byrnes and N. Hurt, *On the moduli of linear dynamical systems,* Advances in Mathematics, Studies in Analyses, Vol. 4, pp. 83-122 (1978).

ASSOCIATION FOR PHYSICAL AND SYSTEMS MATHEMATICS
BROOKLINE, MASSACHUSETTS
U.S.A.

DEPARTMENT OF MATHEMATICS AND STATISTICS
CASE WESTERN RESERVE UNIVERSITY
CLEVELAND, OHIO
U.S.A.

BAYESIAN SYSTEMS IDENTIFICATION IN THE PRESENCE OF SMALL NOISE

Omar Hijab

1. INTRODUCTION. This paper is concerned with the rigorous application of Bayesian estimation ideas to the usual linear state-space model discussed in the literature. Our main result characterizes the conditional statistics of this model explicitly and is nothing more than a careful statement of Bayes' rule. Although known to Fel'dbaum [2], it is surprising that in later years one does not find explicit statements of this result in the literature.

In particular, when there is no parameter uncertainty, our main result states that the Kalman filtering equations characterize the conditional statistics *in the presence of general feedback*, a result not known to hold in this generality until now.

We then use the main result to relate the Bayesian approach to System Identification to the Maximum Likelihood approach to System Identification: We scale the variances of the initial, state and observation noises by a factor ε and study the behavior of both parameter estimates as $\varepsilon \downarrow 0$. We show that although the Maximum Likelihood Estimate is insensitive to the scaling factor ε, *the Bayes Estimate* depends on ε and moreover *converges to the Maximum Likelihood Estimate as* $\varepsilon \downarrow 0$.

2. BAYES' RULE. Consider the linear stochastic system

$$dx = a_\theta x dt + b_\theta u dt + g_\theta d\xi \qquad x(0) = x_\theta^0 \tag{1}$$

$$dy = c_\theta x dt + d\eta \qquad y(0) = 0 \tag{2}$$

where, for each λ in a parameter space Λ, x_λ^0 is normally distributed with mean m_λ^0 and variance σ_λ^0, and $\xi(\cdot)$ and $\eta(\cdot)$ are standard Brownian motions. For notational simplicity we take u, x, and y all to be scalars although all the results that appear here and their proofs remain valid in the general vector case. Here θ is a fixed random variable taking values in the arbitrary parameter (topological) space Λ and represents our uncertainty as to the actual value of the parameter, and a, b, c, and g are arbitrary continuous functions of λ in Λ and $t \geq 0$.

By a *control* we shall mean *any* stochastic process $u(\cdot)$ satisfying

$$E\left(\int_0^t u(s)^2 ds\right) < \infty$$

for all $t \geq 0$. We assume that $\xi(\cdot)$, $\eta(\cdot)$, θ, x_λ^0, λ in Λ are all mutually independent. It is well-known then that (1) has a unique solution $x(\cdot)$ corresponding to any given control $u(\cdot)$. Note however that $x(\cdot)$ need not have any finite moments. In what follows we neither assume nor need the finiteness of any moments of the state process $x(\cdot)$.

Let $G(m,\sigma)(dx)$ denotes the Gaussian distribution of mean m and variance σ on the real line $\mathbf{R}$ and let the a priori probability distribution of θ on Λ be denoted by $\pi^0(d\lambda)$. Given any control $u(\cdot)$, the a posteriori conditional probability distribution $\pi(t,d\lambda)$ of θ given $y(s)$, $0 \leq s \leq t$, is well-defined and

$$\int_\Lambda \phi(\lambda)\pi(t,d\lambda) = E(\phi(\theta)|y(s),\ 0 \leq s \leq t) \quad ,$$

for any bounded continuous function ϕ and Λ. Let $p(t,dx,d\lambda)$ denote the conditional joint probability distribution of $x(t)$ and θ given the observations $y(s)$, $0 \leq s \leq t$, corresponding to a given control:

$$\int_{\mathbf{R}\times\Lambda} \phi(x,\lambda)p(t,dx,d\lambda) = E(\phi(x(t),\theta)|y(s),\ 0 \leq s \leq t) \quad ,$$

for any bounded continuous function ϕ on $\mathbf{R}\times\Lambda$. If $u(\cdot)$ is a control, let $m_\lambda(\cdot)$ and $\sigma_\lambda(\cdot)$ denote the unique solutions of

$$dm_\lambda = (a_\lambda m_\lambda + b_\lambda u)dt + \sigma_\lambda c_\lambda(dy - c_\lambda m_\lambda dt) \qquad m_\lambda(0) = m_\lambda^0 \tag{3}$$

$$\dot{\sigma}_\lambda = 2a_\lambda\sigma_\lambda + g_\lambda^2 - \sigma_\lambda^2 c_\lambda^2 \qquad \sigma_\lambda(0) = \sigma_\lambda^0 \quad . \tag{4}$$

Then for each λ in Λ, $m_\lambda(\cdot)$ and $\sigma_\lambda(\cdot)$ are the Kalman mean and variance corresponding to the control $u(\cdot)$ and parameter λ. In what follows we assume that m_λ^0 and σ_λ^0 are continuous functions of λ in Λ.

Let $\ell_\lambda(t)$ denote the *likelihood* [1], [4] of the parameter λ given the observations $y(s)$, $0 \leq s \leq t$,

$$\ell_\lambda(t) \triangleq \exp\left(\int_0^t c_\lambda(s)m_\lambda(s)dy(s) - \frac{1}{2}\int_0^t (c_\lambda(s)m_\lambda(s))^2 ds\right) \tag{5}$$

up to time t. As far as implementation is concerned, this is just the exponential of the residuals obtained from the Kalman filter "tuned to" the parameter λ. We come now to an important concept. To each control $u(\cdot)$ corresponds uniquely via (1) and (2) an observation process $y(\cdot)$. We say that $u(\cdot)$ is *admissible* iff $u(t)$ depends only on $y(s)$, $0 \leq s \leq t$, for all $t \geq 0$. More precisely, let $\mathfrak{Y}_t^u$ denote the past of the process $y(\cdot)$ up to time t. Then $u(\cdot)$ is *admissible* iff $u(t)$ is $\mathfrak{Y}_t^u$-measurable for all $t \geq 0$. We emphasize that $\mathfrak{Y}_t^u$ may depend on the choice of $u(\cdot)$ and that the dependence of $\mathfrak{Y}_t^u$ on the process $u(\cdot)$ *plays no role* in the proof of Theorem 1 below.

Let $C([0,t])$ denote the space of all observation records $y(s)$, $0 \leq s \leq t$, with Wiener measure imposed on it. Then the admissible control $u(t)$, the Kalman mean $m_\lambda(t)$, the conditional distribution $p(t,dx,d\lambda)$, and the likelihood $\ell_\lambda(t)$ may all (and *will*) be thought of as measurable functions on $C([0,t])$, for all $t \geq 0$.

THEOREM 1. *Fix an admissible control* $u(\cdot)$ *and suppose that*

$$E\left(\int_0^t (c_\theta(s)x(s))^2 ds\right) < \infty \tag{F}$$

for all $t \geq 0$, in addition to the assumptions stated so far. Then the conditional joint probability distribution of $x(t)$ and θ given the observations $y(s)$, $0 \leq s \leq t$, is given by

$$p(t,dx,d\lambda) = \frac{G(m_\lambda(t),\sigma_\lambda(t))(dx) \times \ell_\lambda(t) \times \pi^0(d\lambda)}{\int_\Lambda \ell_\lambda(t) \times \pi^0(d\lambda)} ,$$

almost surely on the space of all observation records $C([0,t])$, for all $t \geq 0$.

With a slight modification of notation, this theorem appears as Theorem 4 in [3]. Note that if Λ consists of one point then we are reduced to a single Kalman filter, i.e. the Kalman filtering equations characterize the conditional statistics provided only that $u(\cdot)$ be an admissible control. Note also that when Λ is compact, condition (F) holds automatically.

As a corollary we see that

$$\pi(t,d\lambda) = \frac{\ell_\lambda(t) \times \pi^0(d\lambda)}{\int_\Lambda \ell_\lambda(t) \times \pi^0(d\lambda)} \tag{6}$$

for all $t \geq 0$.

3. ASYMPTOTICS. Recall the state-space model (1) and (2) and *rescale the variances* be replacing σ^0 by $\varepsilon\sigma^0$, $\xi(\cdot)$ by $\sqrt{\varepsilon}\,\xi(\cdot)$, and $\eta(\cdot)$ by $\sqrt{\varepsilon}\,\eta(\cdot)$. Let the corresponding quantities be denoted by $m_\lambda^\varepsilon(t)$, $\sigma_\lambda^\varepsilon(t)$, $\ell_\lambda^\varepsilon(t)$, and $\pi^\varepsilon(t,d\lambda)$. A quick run through (3) and (4) then reveals that

$$m_\lambda^\varepsilon(t) = m_\lambda(t) \qquad \text{and} \qquad \sigma_\lambda^\varepsilon(t) = \varepsilon\sigma_\lambda(t) \qquad !$$

To see this, note that rescaling $\xi(\cdot)$ is the same as replacing g by $\sqrt{\varepsilon}\,g$ and that rescaling $\eta(\cdot)$ is the same as replacing c by $c/\sqrt{\varepsilon}$ and $y(\cdot)$ by $y(\cdot)/\sqrt{\varepsilon}$. Now (5) says that the likelihood is given by

$$\ell_\lambda^\varepsilon(t) = (\ell_\lambda(t))^{1/\varepsilon}$$

and thus (6) is replaced by

$$\pi^\varepsilon(t,d\lambda) = \frac{(\ell_\lambda(t))^{1/\varepsilon} \times \pi^0(d\lambda)}{\int_\Lambda (\ell_\lambda(t))^{1/\varepsilon} \times \pi^0(d\lambda)} \qquad (6^\varepsilon)$$

Thus *any maximum likelihood estimate will not depend on the scaling factor* ε. The following technical lemma is not necessary in discrete time.

LEMMA. *Assume that* $dc_\lambda(t)/dt$ *exists and is a continuous function of* λ *in* Λ *and* $t \geq 0$ *and fix an admissible control* $u(\cdot)$. *Then* $\ell_\lambda(t)$ *can be defined for* all *observation records* $y(s)$, $0 \leq s \leq t$, *and is a continuous function of* λ *in* Λ, *for each observation record* $y(s)$, $0 \leq s \leq t$.

Proof. The only obstacle to continuity with respect to λ is the presence of the stochastic integrals in (3) and (5). However, because c is differentiable, we can evaluate by parts the stochastic integral appearing in (3) and conclude that $m_\lambda(s)$, $0 \leq s \leq t$, can be defined for each observation record $y(s)$, $0 \leq s \leq t$, and is then a continuous function of λ. Similarly we evaluate by parts (twice) the stochastic integral appearing in (5) and so the result follows.

In what follows we shall work with the version of $\ell_\lambda(t)$ described by the above lemma. Without loss of generality, we may assume that the support of the probability distribution of θ is all of Λ i.e. we may assume that for any open set U in Λ we have $\pi^0(U) > 0$.

DEFINITION. *Given an observation record* $y(s)$, $0 \leq s \leq t$, a parameter $\hat{\lambda}$ in Λ satisfying

$$\ell_\lambda(t) < \ell_{\hat{\lambda}}(t) \quad ,$$

for all λ not equal to $\hat{\lambda}$ in Λ, is called *the maximum likelihood estimate for the observation record* $y(s)$, $0 \leq s \leq t$.

We then have the following.

THEOREM 2. *Let* $y(s)$, $0 \leq s \leq t$, *be given and suppose that the corresponding maximum likelihood estimate* $\hat{\lambda}$ *exists. Then for any bounded function* ϕ, *of compact support, that is continuous at* $\hat{\lambda}$ *we have*

$$\lim_{\varepsilon\downarrow 0} \int_\Lambda \phi(\lambda)\pi^\varepsilon(t,d\lambda) = \phi(\hat\lambda) \quad . \tag{7}$$

Proof. This result is a special case of (far more) general results of Varadhan on the asymptotic evaluation of integrals [5]. For completeness however we include a self-contained proof.

First it is enough to verify (7) for nonnegative ϕ. Second let us verify (7) in the case that ϕ vanishes in a neighborhood U of $\hat\lambda$. In this case the maximum of $\ell_\lambda(t)$ on the support of ϕ is strictly less than $\ell_{\hat\lambda}(t)$ and so

$$\overline{\lim_{\varepsilon\downarrow 0}}\, \varepsilon\log\left(\int_\Lambda \phi(\lambda)(\ell_\lambda(t))^{1/\varepsilon}\, \pi^0(d\lambda)\right) < \log(\ell_{\hat\lambda}(t)) \quad .$$

Since $\ell_\lambda(t)$ is continuous at $\hat\lambda$, for each $\delta > 0$ we can find a neighborhood U of $\hat\lambda$ such that on U, $\ell_\lambda(t)$ is greater than $\ell_{\hat\lambda}(t) - \delta$. Thus

$$\underline{\lim_{\varepsilon\downarrow 0}}\, \varepsilon\log\left(\int_\Lambda 1\ (\ell_\lambda(t))^{1/\varepsilon}\, \pi^0(d\lambda)\right) \geq \log(\ell_{\hat\lambda}(t) - \delta) \quad .$$

Now by combining the above two inequalities with (6^ε) we arrive at

$$\overline{\lim_{\varepsilon\downarrow 0}}\, \varepsilon\log\left(\int_\Lambda \phi(\lambda)\pi^\varepsilon(t,d\ )\right) < 0$$

which implies (7) since in this case $\phi(\hat\lambda) = 0$. Third for any nonnegative ϕ that is continuous at $\hat\lambda$ and any $\delta > 0$, we can find an open neighborhood U of $\hat\lambda$ such that ϕ is within δ of $\phi(\hat\lambda)$ on U. Thus

$$\phi(\hat\lambda) - \delta \leq \underline{\lim_{\varepsilon\downarrow}} \int_U = \underline{\lim_{\varepsilon\downarrow 0}} \int_\Lambda \leq \overline{\lim_{\varepsilon\downarrow 0}} \int_\Lambda = \overline{\lim_{\varepsilon\downarrow 0}} \int_U \leq \varepsilon(\hat\lambda) + \delta \quad ,$$

for all $\delta > 0$, which completes the proof.

Thus the Bayes estimate converges to Maximum Likelihood Estimate as the variances of the impinging noises go to zero.

REFERENCES

1. T. Duncan, *Probability Densitites for Diffusion Processes with Applications to Nonlinear Filtering Theory and Detection Theory*, Ph.D. Dissertation, Stanford University 1967.

2. A. A. Fel'dbaum, *Optimal Control Systems*, Academic Press, New York 1965.

3. O. Hijab, *The adaptive LQG problem*, to appear in the IEEE Trans. Aut. Control, 1982.

4. R. S. Liptser and A. N. Shiryayev, *Statistics of Random Processes*, Vol. I, Springer-Verlag, New York 1977.

5. S. R. S. Varadhan, *Asymptotic probabilities and differential equations*, Comm. Pure Appl. Math., Vol. XIX, 261-286 (1966).

DEPARTMENT OF MATHEMATICS AND STATISTICS
CASE WESTERN RESERVE UNIVERSITY
CLEVELAND, OHIO 44106
U.S.A.

OPTIMAL ADAPTIVE CONTROL OF MARKOVIAN SYSTEMS*

P. R. Kumar

1. INTRODUCTION. We consider a controlled Markov system with complete observations of the state. A cost criterion is given, which is of the long-term average type. The problem we examine is this: if the dynamics of the stochastic system are unknown, and we may assume that such uncertainty in the dynamics is reflected in our lack of knowledge of a parameter describing the dynamics, then how should we choose *adaptive control laws* which will lead to adequate performance?

In what follows we shall provide a brief description of the setup, the adaptive control law proposed and the pertinent results.

2. GENERAL MARKOVIAN SYSTEM SPECIFICATION. Let X be a Polish state space and U a Polish control set. We also have a finite set A representing the possible values of an unknown parameter. We assume that there is a nonnegative Borel measure μ on X such that for each parameter value $\alpha \in A$, the dynamics of the system for this parameter value are specified by the transition probability function

$$\mathrm{Prob}(x_{t+1} \in B | x_t, u_t, \alpha) = \int_B p(x_t, y, u_t, \alpha)\mu(dy) \tag{2.1}$$

Here $B \subseteq X$ is a Borel set, p is a jointly measurable Borel measure density and x_t and u_t denote the state and control respectively at time t. For each parameter value α, we let $\phi(\cdot,\alpha) : X \to U$ be a Borel measurable stationary control law which it would be appropriate to apply to the system if the parameter value was α.

We assume:

(i) $p(x,y,\phi(x,\alpha),\alpha')$ is continuous in (x,y) for every $\alpha,\alpha' \in A$

(ii) there exists a positive integer m and, for every given open set $0 \subseteq X$, a constant $\theta(0) > 0$ such that for every $x \in X$ there is an integer $n(x,0) < m$ such that

*The research reported has been supported by the U.S. Army Research Office Under Contract No. DAAG-29-K-0038.

$$\text{Prob}(x_{t+n} \in O \mid x_t = x,\ \phi(\cdot,\alpha') \text{ is used, } \alpha \text{ is the model}) \geq \theta$$
$$\text{for all } \alpha,\alpha' \in A \quad . \tag{2.2}$$

(iii) $p(x,y,\phi(x,\alpha'),\alpha) > 0$ if and only if it is so for all $\alpha \in A$. Also

$$\int_X p(x,y,\phi(x,\alpha'),\alpha'')\ln \frac{p(x,y,\phi(x,\alpha'),\alpha)}{p(x,y,\phi(x,\alpha'),\alpha'')}\mu(dy)$$

is a bounded continuous function of x for every $\alpha,\alpha',\alpha'' \in A$.

(iv) there exists a $q(y) \geq p((x,y),\phi(x,\alpha),\alpha')$ with $\int_X q(y)\mu(dy) < \infty$.

(v) $c : X \times X \times U \to \mathbb{R}$ is a bounded Borel measurable cost function with $c(x,y,\phi(x,\alpha))$ continuous in X for each fixed y,α.

It is appropriate to mention here that an example of a system satisfying (2.i-2.iv) is the discrete time scalar system

$$x_{t+1} = f(x_t,u_t,\alpha) + \sigma(x_t,u_t,\alpha)w_t \tag{2.3}$$

where f and σ are bounded continuous functions with $|\sigma| \geq \epsilon > 0$ uniformly and $\{w_t\}$ is a sequence of independent identically distributed $N(0,1)$ random variables. $\{\phi(\cdot,\alpha)\}$ can be any set of continuous functions.

In [1] it is shown that there exists a $J(\phi_\alpha,\alpha')$ which is the almost sure long-term average cost:

$$\lim_{t\to\infty} \frac{1}{t} \sum_{s=0}^{t-1} c(x_s,x_{s+1},u_s)$$

when the control law ϕ_α is used on the model α. It is also shown in [1, Theorem 5] that

(i) there exists an unique $\pi(\cdot,\phi_\alpha,\alpha')$ for which

$$\int_X \pi(x,\phi_\alpha,\alpha')\mu(dx) = 1$$

and

$$\int_X \pi(x,\phi_\alpha,\alpha')p(x,y,\phi_\alpha(x),\alpha')\mu(dx) = \pi(y,\phi_\alpha,\alpha') \tag{2.4}$$

(ii) there exists a bounded continuous $w(\cdot,\phi_\alpha,\alpha')$ such that

$$w(x,\phi_\alpha,\alpha') + \int_X \pi(y,\phi_\alpha,\alpha')\int_X p(y,z,\phi_\alpha(y),\alpha')c(y,z,\phi_\alpha(y))\mu(dz)\mu(dy)$$

$$= \int_X p(x,y,\phi_\alpha(x),\alpha')c(x,y,\phi_\alpha(x))\mu(dy) + \int_X p(x,y,\phi_\alpha(x),\alpha')$$

$$\cdot\, w(y,\phi_\alpha,\alpha')\mu(dy) \qquad \text{for every} \quad x \in X \quad \text{and} \quad \alpha,\alpha' \in A.$$

Regarding the choice of the control laws $\{\phi_\alpha\,;\alpha \in A\}$ we assume that

$$J(\phi_\alpha,\alpha) + w(x,\phi_\alpha,\alpha) \le \int_X p(x,y,\phi(x,\alpha'),\alpha)[c(x,y,\phi(x,\alpha')) + w(y,\phi_\alpha,\alpha)]\mu(dy) \tag{2.5}$$

3. SPECIFICATION OF ADAPTIVE CONTROL LAW. To fix notation, let α^0 be the true value of the parameter. We would like to design an adaptive control law which chooses $u_t = u_t(x_0,u_0,x_1,u_1,\ldots,x_t)$ as a function of previous history but does *not* depend on α^0.

To define the adaptive control law we prechoose

(i) a function $J(\alpha) = h(J(\phi_\alpha,\alpha))$ where h is any positive, strictly monotone increasing function. (3.1)

(ii) a positive function $o(t)$ satisfying $\lim_{t\to\infty} o(t) = +\infty$ while

$$\lim_{t\to\infty} \frac{o(t)}{t} = 0 \quad .$$

The adaptive control law consists of making an "estimate" of the unknown parameter according to:

$$\hat{\alpha}_t = \arg\max_{\alpha\in A} J(\alpha)^{-o(t)} \sum_{s=0}^{t-1} p(x_s,x_{s+1},u_s,\alpha) \qquad t = 0,m,2m,\ldots,km,\ldots$$

$$= \hat{\alpha}_{km} \quad \text{for} \quad km \le t < (k+1)m \quad . \tag{3.2}$$

Then at time t it consists of using the control input:

$$u_t = \phi(x_t,\hat{\alpha}_t) \quad . \tag{3.3}$$

This adaptive control law (3.2, 3.3) is of the same type as that first proposed in [2] for the problem of adaptively controlling a finite-state, finite-control Markovian system. Here the restrictions of finiteness on the state and control spaces are eliminated and we allow systems with Polish state and control spaces. A specific example of the type of system which satisfies our hypotheses and is also of interest is that described by (2.3).

4. THE MAIN RESULTS. Let $\Phi := \{\phi_\alpha : J(\phi_\alpha, \alpha^0) = J(\phi_{\alpha^0}, \alpha^0)\}$ be the set of control laws which are optimal for α^0 in the class considered. Note also that $J(\phi_{\alpha^0}, \alpha^0)$ is the minimum cost achievable for the true system even if one knew the value of the true parameter α^0 at the start.

For the adaptive control law (3.2), (3.3) we have the following results.

THEOREM.

(i) $$\lim_{t\to\infty} \frac{1}{t} \sum_{s=0}^{t-1} 1(\phi_{\hat{\alpha}_s} \in \Phi) = 1 \qquad \text{a.s.}$$

where $1(\cdot)$ *is the indicator function of the event described in parenthesis.*

(ii) $$\lim_{t\to\infty} \frac{1}{t} \sum_{s=0}^{t-1} c(x_s, x_{s+1}, \phi(x_s, \hat{\alpha}_s)) = J(\phi_{\alpha^0}, \alpha^0) \qquad \text{a.s.}$$

Proof. See [1].

The first assertion states that almost surely the adaptive control law Cesaro-converges to the set of optimal control laws. The second assertion states that the actual long-term average cost incurred by the adaptive control law is optimal (almost surely) and cannot be bettered even if one knew the value of the unknown parameter at the start.

REFERENCES

1. P. R. Kumar, *Simultaneous identification and adaptive control of unknown systems over finite parameter sets*, Math. Research Report No. 81-5, University of Maryland Baltimore County, June 1981.

2. P. R. Kumar and A. Becker, *A new family of optimal adaptive controllers for Markov chains*, IEEE Transactions on Automatic Control, Vol. AC-27, pp. 137-146, February 1982.

DEPARTMENT OF MATHEMATICS
UNIVERSITY OF MARYLAND BALTIMORE COUNTY
5401 WILKENS AVENUE
BALTIMORE, MARYLAND 21228
U.S.A.

ALGEBRAIC GEOMETRY AND THE BUSINESS CYCLE

Steven E. Landsburg

1. THE BUSINESS CYCLE. While few economists could confidently offer a precise definition of "The Business Cycle," fewer still would deny that it exists. By 1913, the National Bureau of Economic Research under the direction of Wesley Mitchell had begun to document the persistence of comovements in certain aggregative economic time series. They found that such quantities as output, profits, interest rates, the money supply, the price level and the level of employment tend to "move together"--and that they repeatedly "move together" in the same general way.[1] This conclusion--fortified by reams of empirical evidence published by the Bureau in 1951 [M] and later codified into the slogan that "business cycles are all alike" [L1] makes modeling the business cycle an interesting and potentially tractable problem for the economic theorist.

Of course, the delineation of a persistent relationship between variables which are at least partially subject to government control (such as the money supply, and through it the price level) on the one hand, and variables such as output and employment on the other, is not purely of theoretical interest--as anybody with knowledge of the extended business cycle trough of the 1930s can testify. One hopes that it would lead to wiser public policy, and (among other things) to the prevention or mitigation of events like the Great Depression.

2. KEYNESIAN MODELS AND THEIR ECONOMETRICS. These considerations inspired Keynes's <u>General Theory of Employment, Interest and Money</u>, the first abstract economic model which explicitly attempted to account for cyclical phenomena such as Mitchell observed. (For a fascinating, if partisan, account of the history of business cycle modeling in general, see Lucas [L2].)

A primary task of any model of the business cycle is to explain the empirical correlation between employment and the rate of change in the price level, which persisted at least until the 1970s. (Later on, we will raise the question of why the 1970s were so different.) This correlation is often referred to as the "inflation-unemployment trade-off" represented by a downward-sloping "Phillips Curve" relating unemployment on the horizontal axis to the inflation rate on the vertical.

The question, then, is "Why is there a Phillips Curve"? The Keynesian answer, in part, is that wages are "sticky"--that they fail to adjust quickly to the new equilibrium level of prices. Thus, for example, if the general price level were to fall by a third (as it did in the early thirties) while wages, being "sticky," failed to adjust accordingly, then employers would not be prepared to hire as much labor as they were previously, and many workers would become "involuntary unemployed."

Note that this is very much a "non-equilibrium" theory in the sense that wages fail to adjust to the point where the quantity of labor demanded is equal to the quantity supplied.

In order to convert such a model (or any economic model) into a form which is useful for making predictions, one writes down a system of equations that captures the relationships which the model says "ought" to exist, assumes that the equations take a certain form (usually linear) and uses observations of the variables to estimate the parameters of the equations.

This, of course, raises an "identification problem" in the sense of the econometrician: several choices of parameters may be observationally equivalent, so that no data are capable of distinguishing between them.

Rather than illustrate these points with a complicated macroeconometric model, let me recall the world's most elementary example of an identification problem: the "Marshallian Cross" of supply and demand. The relevant variables are price (denoted P), quantity (denoted Q) and two random variables u and v. The model postulates a supply curve and a demand curve:

$$P = a_0 + a_1 Q + u$$

$$Q = b_0 + b_1 P + v$$

which jointly determine P and Q as functions of u and v. (Presumably a_1 is positive and b_1 negative.) Then, even given full information about the distributions of u and v and given full information about the joint distribution of P and Q that is derived from the model, one still can not determine a_0, b_0, a_1 and b_1 uniquely.

However, for purposes of policy evaluation it is clearly desirable to know these parameters. For example, we know from economic theory that an excise tax of \$T per unit sold will have the effect of replacing a_0 by $a_0 + T$ while leaving the other parameters unchanged. Thus if we wished to know in advance the effects of an excise tax on the equilibrium distribution of P and Q, it would suffice to know the values of a_0, b_0, a_1 and b_1.

One way to deal with such an identification problem, as everyone knows, is to introduce a combination of exogenous variables and exclusion restrictions. Thus, suppose we let Z denote the price of some raw material and Y average household income, and we rewrite the model equations as

$$P = a_0 + a_1Q + a_2Z + a_3Y + u$$

$$Q = b_0 + b_1Q + b_2Z + b_3Y + v \quad .$$

If we take Z and Y to be exogenous (i.e. determined by factors outside the model) *and* if we specify $a_3 = b_2 = 0$ (which has some intuitive plausibility) then the model becomes identified. (The restriction $a_3 = b_2 = 0$ is an example of what I mean by an "exclusion restriction.")

In Keynesian macroeconometrics, there are several large (and empirically quite successful) models, usually identified through a combination of three techniques:

(a) the declaration of certain variables as exogenous;

(b) exclusion restrictions; i.e. a priori requirements that certain parameters or linear combinations of parameters be set equal to zero; and

(c) assumptions about the covariances of the error terms.

Economists refer to such identifying restrictions collectively as being of the "Cowles Commission type." The standard reference for this classical approach is Fisher [F].

However, before immersing himself in this literature (or congratulating himself on his familiarity with it) the reader should be made aware of the fly in the ointment. Two closely related theoretical developments which we shall presently explore (and considerable empirical evidence as well) lead to the conclusion that the Keynesian approach to model specification and identification can not in general yield information which is of any use in evaluating the potential effects of alternative economic policies. Insofar as the latter is a major goal of economics, this approach must be judged a failure.

One of these developments was Lucas's "Critique" of econometric policy evaluation [L3] which argued that the structural parameters which are usually estimated in Keynesian macroeconomic models (and in many other economic models as well) are not invariant under changes of policy. Thus if their values are estimated under a fixed policy regime they can not be used to make forecasts about the effects of hypothetical alternative regimes.

The other development, inspired by Friedman [Fr] and Phelps [P], was a theory of unemployment which in its conception, in its predictions, in its implications for economic policy, and in its implications for econometric practice, differs radically from those of Keynes and the Keynesians. The new school of economists who formalized and expounded this theory (most notably Lucas in [L4] and [L5]) have stressed all of these differences, but one in particular will concern us here: identifying restrictions of the Cowles Commission type are fundamentally inconsistent with the logic of these models.

I want to indicate in a very superficial way just what that logic is, before returning to Lucas's Critique.

3. EQUILIBRIUM MODELS OF THE BUSINESS CYCLE. In 1968, in his Presidential address to the American Economic Assocation, Milton Friedman challenged the received wisdom about the origins of the Phillips Curve. The alternative he offered suggested that the observed "inflation-unemployment trade-off" could in fact be characteristic of an equilibrium state in which all individuals behave optimally. This is in stark contrast with the Keynesian view.

Friedman (and similarly Philps [P]) told the following story:

> In equilibrium, fully anticipated changes in the price level should not have any real effects at all. If all prices, all wages, all money holdings, etc. were to double tomorrow, nothing would change except for the numbers on the bills we carry. If it were known in advance that all of these things were going to happen every Wednesday, this fact would be reflected in the "nominal" terms of contracts entered into, but it would not affect the "real" terms--number of hours worked, etc. Indeed, why should it? If we all agreed to stop measuring carpet in square yards and start measuring it in square feet, we would not expect any change in the actual amount of carpeting bought and sold.
>
> (I should note that this story is grossly oversimplified, and rests, among other things, on the ability of the monetary authority to magically make all money balances--including that stack of $20 bills you keep hidden away under your mattress--grow at the same rate as prices and wages. But even fables have morals, so let us proceed.)
>
> Now imagine an unemployed worker. The reason he is unemployed is not that there are no jobs available to him, but that they are only available at wages lower than he is willing to accept. In fact, the highest wage offer he has received is for $8,000 a year, whereas he is not willing to work for less than $10,000.
>
> One night, while our worker is sleeping, all prices and all wages double. He is awakened the next morning by a telephone call from an employer who says, "I am now prepared to offer you an annual salary of $16,000." Of course, $16,000 today will only buy what $8,000 would have bought yesterday, so the worker, if his tastes and constraints are unchanged, and if he is rational and fully informed, will still not accept the position.
>
> But what if he is *not* fully informed? What if he went to sleep unaware of the changes which were to take place in the course of the night, and having just been awakened by a telehpone call is still unaware of them? If that case he will accept the job, convinced that he will be earning far more than his minimum requirement of $10,000.

> Now, after a day on the job, our hero is likely to stop at the supermarket to indulge the temptations of his new economic status. When he sees the prices on the items, he will recognize himself to be the victim of a cruel hoax, and begin the mental task of composing a letter of resignation.

This, in its essence, is the intuitive story that Friedman told in 1968. Like the Keynesian models, it predicts a Phillips Curve. Unlike those models, it predicts that the corresponding trade-off is ephemeral. (If Friedman's story is correct, then sustained high inflation can not produce sustained high levels of employment--because only insofar as inflation is unexpected does it affect employment, and sustained high inflation quickly becomes expected high inflation.)[2]

In 1972 [L4] and 1975 [L5], Lucas showed by example that it is possible to construct a complex model economy which is guided by this intuition and which exhibits a business cycle which Wesley Mitchell would have been hard put to distinguish from the Real McCoy. In these models, individuals face a "signal-processing" problem in which they are attempting to estimate the "real" prices of various products on the basis of their "dollar" prices and incomplete information about the general price level. In so doing, they have an incentive to use all available information in forming their estimates. This means that the value of any variable which can be observed by an agent will affect his assessment of the state of the economy and consequently affect his actions--and through them the values of other variables. This imposes a very complicated structure on the relationships between the variables which is incompatible with identifying restrictions of the Cowles Commission type.

Such models as Lucas's are called "natural rate" models because they predict that in the long run the rate of unemployment is determined by "real" as opposed to "monetary" factors. In this and in their other predictions they confirm the vision of Friedman and Phelps. In particular, they are models in which "money is neutral" in the sense that fully anticipated changes in the average rate of inflation will have no effects on real variables. (This is not true, however, of higher moments. In [La] I investigated the consequences of a fully anticipated change in the *variance* of the rate of inflation, and it is not neutral.) In other words, changes in the average rate of inflation can affect unemployment only when they are unanticipated. Barro [B] has recently tested this prediction and found it to be consistent with the data.

It is this last prediction which is most strikingly at odds with Keynesian theory. In 1968, Friedman argued forcefully that high inflation and high unemployment could coexist as long as the inflation was anticipated. All Keynesian theorists disagreed. The decade of the 1970s constituted what I have heard Bob Lucas call a "very costly but very conclusive experiment," and the Keynesians have some explaining to do.[3]

4. LUCAS'S CRITIQUE. Lucas's Critique of econometric policy evaluation [L3] called the attention of the economics profession to a glaring but incredibly widespread misconception about the uses of econometric models. Often, the functions that are estimated in these models describe the behavior of classes of individuals. This behavior consists in part of reactions to government policies. If the policies change, then so will the behavior being described.

Yet, until Lucas's paper appeared in 1976 (and in some circles even today) it was routine econometric practice to ignore this effect. When the goal is to predict the future of the economy under a fixed policy regime, the effect is non-existent. But when the goal is to predict that future under a hypothetical alternative regime, the effect is devastating.

Returning to the example of the "Marshallian Cross" with an excise tax, let P_s be the price received by suppliers, P_d the price paid by demanders, Q the quantity sold, and T the amount of the tax. The model consists of a supply curve, a demand curve, and an accounting identity:

$$P_s = a_0 + a_1 Q$$

$$Q = b_0 + b_1 P_d$$

$$P_d = P_s + T \quad .$$

(Throwing in stochastic terms would complicate matters slightly without affecting the basic analysis.)

If T is exogenous, the model is identified, and the paramters a_0, a_1, b_0 and b_1 can be calculated provided that T has taken on two distinct values in the past (and assuming that P_s, P_d and Q are always observable). Alternatively, if T has not varied, the model may have been identified by adding additional variables as in Section 2. One way or another, suppose that a_0, a_1, b_0 and b_1 are known.

Then it is possible to calcualte the effect of raising T from \$0 per item to \$1 per item. The quantity sold, for example, will fall from $(b_0 + a_0 b_1)/(1 - a_1 b_1)$ to $(b_0 + a_0 b_1 + b_1)/(1 - a_1 b_1)$.

But this prediction is far too general to stand any chance of always being correct. A \$1 tax which is perceived as permanent may indeed have the effect predicted by the model. A \$1 tax which is expected to be in effect only until it is declared unconstitutional a week from Wednesday will have a radically different effect. A \$1 tax on hamburgers which is in effect only on even-numbered days will not reduce the quantity sold on even-numbered days to $(b_0 + a_0 b_1 + b_1)/(1 - a_1 b_1)$; it is much more likely to reduce the quantity sold on even-numbered days to zero (and to multiply the quantity sold on odd-numbered days by two).

In other words: public perceptions of the future course of government policy affect supply and demand curves today. The parameters a_0, a_1, b_0 and b_1, even if known with certainty, can not be used to predict the effects of any hypothetical change in tax policy because the parameters themselves are not invariant under changes in perceived policy--which in turn is presumably not invariant under changes in the policy itself.[4]

The inadequacy of this model does not reside in its failure to incorporate dynamics. If we inserted lagged values of prices, quantities and taxes in the supply and demand equations, we would have more parameters to estimate, but these parameters would still be derived from public perceptions of the process generating the values of T. A change in that process (if it led to a change in the way the process was perceived) would change those parameters in a way which the econometrician could not foresee.

* * *

The Lucas Critique and the natural rate hypothesis are so closely linked conceptually and historically that they are hard to disentangle. They both underscore the importance of modeling the formation of expectations. They both derive from the principle that people are not robots--that their behavior changes when their environment does.[5]

To show how Lucas's criticism applies to a simple version of a natural rate model, I have lifted an example directly from his original paper:

We suppose that there are N physically separated markets. Everyone knows the past history of every market but cannot make contemporaneous observations of any market but his own.

Suppose that:

y_{it} = log (output in market i at time t)

p_{it} = log (price of output in market i at time t)

p_t = log (average economy-wide price of output at time t)

p^e_{it} = estimate of p_t in market i at time t

z_{it} are independently and identically distributed random variables with mean 0

ε_t are independently and identically distributed random variables with mean π .

Assume that the laws of motion of the economy are given by:

$$y_{it} = \beta(p_{it} - p^e_{it})$$

$$p_{it} = p_t + z_{it}$$

$$p_t = p_{t-1} + \varepsilon_t \quad .$$

Then optimal forecasting on the part of agents requires:

$$p_{it}^{e} = (1-\theta)p_{it} + \theta(p_{t-1} + \pi)$$

where θ is a constant depending on the variances of p_t and z_{it}. (These variances are assumed known to all.)

Solving, adding up, and invoking the Law of Large Numbers, we get total output:

$$y_t \equiv \sum_{i=1}^{N} y_{it} = \theta\beta(p_t - p_{t-1}) - \theta\beta\pi \quad . \tag{*}$$

This economy will exhibit a Phillips Curve. In periods of high inflation ($p_t - p_{t-1}$ large) output will be high. In periods of low inflation, output will be low. An econometrician regressing output (y_t) on inflation ($p_t - p_{t-1}$) will estimate a positive slope coefficient $\theta\beta$ and a constant term $\theta\beta\pi$. (And he will consistently make accurate predictions of y_t as a function of $p_t - p_{t-1}$.)

Now suppose that it is proposed to take advantage of this trade-off by instituting a policy of sustained high inflation. *If* we assume that the behavioral equation (*) will be remain in effect under the new policy (the traditional econometric practice!) then we will predict sustained high levels of output. But in reality, this sustained high inflation (which amounts to a rise in the expected value of ε_t --that is, a rise in π) will lower the constant term of (*) by *exactly* the average increase in the linear term, and so will have no effect on average output!

5. RATIONAL EXPECTATIONS. Can a policy-maker with control over the average rate of inflation exploit the Phillips Curve trade-off to increase output? The answer is: it depends. In the model just described, the answer is "no." But let us modify that model just slightly by introducing a new variable π_0 which represents the "public perception" of π. (Recall that π is the average rate of inflation.)

The equation (*) will be replaced by a new expression for total output:

$$y_t \equiv \sum_{i=1}^{N} y_{it} = \beta\theta(p_t - p_{t-1}) - \beta\theta\pi_0 \quad . \tag{**}$$

A change in π (say to π') will raise the linear term on average by $\beta\theta(\pi' - \pi)$ and lower the constant term by $\beta\theta(\pi_0' - \pi)$ where π_0' is the public perception of π' after the policy change.[6]

Keynesian macroeconometric models implicitly make the extreme assumption that $\pi_0' = \pi_0$, i.e. that the parameters of the structural equation (*) (in this case the constant term) are independent of the policy change. The equally extreme assumption of the Lucas model is that $\pi_0 = \pi$ and $\pi_0' = \pi'$, i.e. that the public always knows the average rate of inflation perfectly (although the driving force behind the model is that its knowledge of the *current* rate of inflation is always *im*perfect). In either case we have the same Phillips Curve, but in one case it is exploitable $(\beta\theta(\pi_0' - \pi_0) = 0)$ and in the other it is not $(\beta\theta(\pi' - \pi) = \beta\theta(\pi_0' - \pi_0))$.

Now to say that two assumptions are equally extreme is not to say that they are equally objectionable. The successful branches of economics--those in which there is widespread professional consensus supported by empirical documentation--are those which take seriously the notion that an economic agent is somebody who optimizes something. If we are going to incorporate this view into our thinking about business cycles, we are going to be forced to admit that agents who learn about the features of the stochastic environment are going to be better at optimizing than agents who do not. The extreme assumption that agents are entirely oblivious to changes in the environment (in the context of our example, that $\pi_0' = \pi_0$) is simply untenable given this point of view.

Of course, our unwillingness to embrace one extreme need not force us into the arms of the other. (I will, however, argue shortly that in this case there may be no reasonable middle ground, and that Lucas's extreme position is in any event not an unattractive one.) Be that as it may, the moral of the Critique is that in a world of optimizing agents it is certainly necessary to have *some* explicit model of expectations formation in order to evaluate policies.

To put this another way, we may view the economy as a *system*, described by laws of motion:

$$x_{t+1} = f(x_t, u_t^{(1)}, u_t^{(2)}, \ldots, u_t^{(N)}, z_t)$$

$$y_t = g(x_t)$$

where x_t is a "state" vector of variables which may be unobservable to the econmetrician and/or the agent, y_t is a vector of "outputs," $u_t^{(i)}$ is a "control" vector representing decisions made at time t by the i-th agent in the economy, and z_t is a vector of variables generated by stochastic processes (such as the price level).

The i-th agent chooses $u_t^{(i)}$ so as to maximize the expected value of an objective function

$$v_i(u_t^{(i)}, u_{t+1}^{(i)}, \ldots, x_{t+1}, x_{t+2}, \ldots)$$

where the expected value is conditional on all information available to agent i at time t (call this set I_{it}) and on the knowledge that for $p > 0$, u_{t+p} will be selected according to the same criterion.

Now the point is that this problem is not well-posed. "Maximizing the expected value of v_i" is a meaningless phrase unless it is made contingent on some belief about the processes generating z_t and $u_t^{(j)}$ for $j \neq i$. Given beliefs about those processes, the optimization problem is well-posed and its solution determines $u_t^{(i)}$ as a function of I_{it}. Let us write this solution as:

$$u_t^{(i)} = h_i(I_{it}) \quad .$$

In this context, Lucas's Critique comes down to these observations: (1) Knowledge of the h_i and of the process generating z_t is certainly sufficient to predict future values of y_t. (2) A change in the process generating z_t (e.g. a change in government policy)--assuming agents are aware of it--will change the optimization problems being solved and so change the solutions h_i. (3) Therefore it is incorrect under these circumstances to predict future values of y_t by inserting a new z-process into the old functions h_i. This incorrect procedure is the basis for policy evaluation in all Keynesian macroeconometric models. (4) Not only will the h_i change when policy changes, but they will change in a way that can not be prediced on the basis of knowledge of the current h_i alone.

So: In order to predict how changes in the z-process will affect outputs, we need to know how they will affect the h_i. The first step in solving this problem is to make an explicit modeling assumption about how changes in the process are perceived by the public.

We have discussed two "extreme" solutions to this modeling problem. The first, that agents are entirely unaware of policy changes, is unacceptable. The second, that agents always know the policy processes precisely, is known as the hypothesis of "rational expectations."

The rational expectations hypothesis is currently much in vogue and much under attack. Many critics raise the objection that it is clearly untrue. But this does not speak to the question of whether or not it is a useful approximation to the truth. In fact, as an approximation it may be not only useful but even reasonably close. Optimizing agents read newspapers and keep their eyes and ears open; in a democratic society, changes in government policy are rarely kept secret for long.

The reasons for accepting the rational expectations hypothesis are that it is conceptually simple, that it is consistent in spirit with the modeling hypotheses that have been found to be successful in other branches of economics, that models which use it are capable of generating "realistic" business cycles (it plays a central role in the models of [L4] and [L5], for example); that it is consistent with the otherwise "anomalous" experience of the 1970s; and that nobody has suggested an alternative with all of these properties.

You don't have to like rational expectations to like the natural rate hypothesis, and you don't have to like rational expectations to buy Lucas's Critique. But if you accept the general circle of ideas surrounding equilibrium business cycle theory, then you must accept the need for *some* explicit modeling hypothesis about expectations. I know of no alternative which is as attractive in as many ways as the hypothesis that expectations are rational.

After this defense, I now come around to admitting that "assuming rational expectations" actually asks for a bit more credulity than I have so far let on. The i-th agent in our model chooses his optimizing strategy $u_t^{(i)} = h_i(I_{it})$ subject to beliefs not only about z but about the other h_j and about f and g as well. I will take the "assumption of rational expectations" to mean that he has correct beliefs about all of these functions as well. (Thus each agent is assumed to know in advance all of the parameters which are to be estimated by the econometrician. In that sense, rational expectations is the economics of Gerald Ford, who was fond of saying that the average housewife in a supermarket knew more about the workings of the economy than the entire economics profession.)

These hypotheses, together with some modeling assumptions about the objective functions v_i, the laws of motion f and g, the information sets I_{it}, and the process z, severely limits the set of n-tuples (h_i) that can occur.

Explicitly: Given v_i, f, g, I_i, and z, define a *rational expectations equilibrium* to be a collection of functions h_i such that for each i, the feedback rule $u_t^{(i)} = h_i(I_{it})$ maximizes the objective function:

$$E_{it}(v_i(h_1(I_{1t}),h_2(I_{2t}),\ldots,h_{i-1}(I_{i-1,t}),u_t^{(i)},h_{i+1}(I_{i+1,t}),\ldots,h_N(I_{N,t})),$$
$$x_{t+1},x_{t+2},\ldots)$$

where E_{it} denotes the expectation conditional on the information in I_{it}.

Now we restrict the forms of v_i, f, g, I_i, and z (e.g. v_i quadratic, f and g linear, I_{it} = the union of all information with time subscript less than t with $p(z_t)$ where p is a linear function of small rank, z_t of the form $\sum_{k=1}^{m}\theta_k z_{t-k} + b$ where the θ_k are constants and b is white noise). We may make further restrictions as well, such as that all the v_i be identical. Given such restrictions, the requirement that the h_i constitute a rational expectations equilibrium is a very powerful one.

In fact, it is so powerful that it may serve to identify the model. Given observations of $u^{(1)},\ldots,u^{(N)}$, y and z over time, we wish to estimate the model

$$x_{t+1} = f(x_t, u_t^{(1)}, \ldots, u_t^{(N)}, z_t)$$

$$y_t = g(x_t)$$

$$u_t^{(i)} = h_i(I_{it})$$

$$z_t = j(z_{t-1}, z_{t-2}, \ldots, u) \quad .$$

With modeling hypotheses such as those in the last paragraph, the optimal h_i turn out to be linear, so our model is

$$x_{t+1} = \alpha x_t + \sum_{j=1}^{N} \beta_j u_t^{(j)} + \gamma z_t + a_1$$

$$y_t = \delta x_t + a_2$$

$$u_t^{(i)} = \sum_{k=1}^{m} \varepsilon_k y_{t-k} + \sum_{k=1}^{m} \sum_{j=1}^{N} \zeta_{kj} u_{t-k}^{(j)} + \eta p(z_t) + a_3$$

$$z_t = \sum_{k=1}^{m} \theta_k z_{t-k} + b$$

with a_1, a_2 and a_3 interpreted as measurement errors and b as fundamental white noise. (The Greek letters represent linear maps of the appropriate dimensions.)

The problem is to estimate the parameters denoted by Greek letters from observed data. The preliminary problem is to identify the model--to make assumptions that would uniquely determine those parameters if the true joint distribution of the observables were known. I have argued the unsuitability of the Cowles Commission techniques of a priori exogeneity, exclusion restrictions, and covariance restrictions. The alternative is to assume that the h_i form a rational expectations equilibrium for f, g, z, and *some* choices of v_i. This imposes explicit (and often very complicated) relationships which the parameters must satisfy. In some cases it imposes enough relationships to achieve identification. The parameters can then be estimated, for example, by the method of maximum likelihood.

Now where does this get us? There are two reasons to build macroeconomic models. The first is to predict the future of the economy under a fixed set of policies and the second is to evaluate possible alternative policies. Identifying and estimating the h_i can only help with the first of these goals. In this sense, our achievement so far is purely an aesthetic one. We have a new way to identify the h_i, and our new way is more compatible with economic theory than the old way. But the fact is that the old way performed impressively well (during periods of stable public policy). Given what we have done so far, the rational expectations approach is more satisfying to an economic theorist, but not more useful to anyone.

In order to *make* it useful, we need only travel one more step. One can ask for circumstances in which the h_i, f, g and z (all of which have been estimated) determine the agents' objective functions v_i uniquely. (In order for this question to even make much sense, it is imperative that the h_i have been identified in some way which takes account of the objective functions' existence.) In those fortunate instances, we can calculate the v_i that are being maximized. Then if we wish to know the effect of a policy change (a new z-process), we have enough information to work out the new rational expectations equilibrium--the new set of functions h_i which Lucas warns us will come to be. Having determined the h_i that are consistent with the new z-process, we can plug everything into our model and make predictions.

This entire process has been carried out in a sophisticated context by Hansen and Sargent [HS]. Sargent [S] provides a beautiful and instructive example of the same approach. In both cases, rational expectations serves both to identify the h_i and to determine the v_i uniquely. (This determination of the v_i is simply a new identification problem, at a deeper level.)

Before remarking on some technical problems associated with this program, let me summarize our goals and our accomplishments:

(1) We begin with two intimately related challenges to Keynesian macroeconomic practice: the natural rate hypothesis and the Lucas Critique.

(2) The natural rate hypothesis invalidates the Cowles Commission techniques for identifying the "control laws" h_i.

(3) The Lucas Critique implies that even if they are correctly identified and estimated, the h_i will not help us evaluate the effects of a policy change.

(4) The hypothesis of rational expectations is a source of identifying restrictions consistent with the natural rate hypothesis and so can serve to meet the first of these challenges.

(5) Identifying the h_i via rational expectations involves the introduction of objective functions v_i. When the h_i are estimated, the v_i are often then determined uniquely (that is, they are identified as well). Knowledge of the v_i is sufficient to predict the effects of policy changes. Therefore in this case, the challenge of the Lucas Critique is met as well.

6. ALGEBRAIC GEOMETRY AND THE BUSINESS CYCLE. The most difficult step in this program is the explicit computation of rational expectations equilibria. This has only been accomplished in a few very special cases. New techniques in this area are much in demand.

Several other technical difficulties--major in some models and minor in others--arise as well. These include determining when the rational expectations restrictions are sufficient to identify the control laws, determining when they are sufficient to identify the objective functions, and developing computationally efficient methods of parameter estimation in the rational expectations context.

Now, people like Chris Byrnes [By] and Roger Brockett [Br] have stressed the importance of viewing problems of identification and parameter estimation in terms of spaces of moduli for systems and transfer functions. I want to argue that this geometric point of view may be especially appropriate to the problems raised by rational expectations.

Let us restate the problem in geometric terms. For each agent i, we have a space of possible objective functions V_i, and a space of possible control laws H_i. We fix the information sets available to agents once for all, and suppress them in our notation. The economy is governed by laws of motion drawn from a space of systems S and a stochastic process from a space of processes Z.

Write V for $V_1 \times V_2 \times \ldots \times V_N$ and H for $H_1 \times H_2 \times \ldots \times H_N$. Then an *economy* is an element of $E = V \times S \times Z$. To such an economy (v,s,z) is associated a *rational expectations equilibrium* $\rho(v,s,z) = h \in H$. (I am finessing the difficult issues of existence and uniqueness!) Define the *working economy* $\omega(v,s,z)$ to be $(\rho(v,s,z),s,z) \in H \times S \times Z$. Then the set W of working economies is the image of the mapping $\omega : E \to H \times S \times Z$. Estimating a model subject to rational expectations amounts to picking a point in W.

We also have a map $\tau : (H \times S \times Z) \to T$ which associates to any (h,s,z) certain observable behavior (so that T is something like a space of joint distributions for observable variables). To say that the problem of estimating the control laws is identified is to say that if we say that if we restrict the mapping τ to $W \subset (H \times S \times Z)$, it becomes a one-to-one mapping.[7] To say that the problem of estimating the objective functions is identified is to say that the mapping ω is one-to-one.

In order to do any practical econometrics, it is necessary to have a way to refer to a point $w \in W$; the space W has to be parameterized. In the example of the last section, W was parameterized for free: we chose V, S, and Z in such a way that H could be taken to be a space of linear functions and hence in a natural way a Euclidean space. Because S and Z were also Euclidean spaces, $W \subset (H \times S \times Z)$ inherited natural coordinates.[8]

However, from the point of view of computational efficiency, these natural coordinates are very unlikely to be the right coordinates. For one thing, there are too many of them. The dimension of W tends to be much smaller than that of $H \times S \times Z$, so that the "natural" choice has a lot of flab to it.

When $\omega : E \to W$ is one-to-one (or generically one-to-one) the natural coordinates on E can be used to parameterize W efficiently. This is precisely what Sargent [S] and Hansen and Sargent [HS] do. But in the case in which E is also of much higher dimension than W (and we will discuss shortly an interesting case in which this occurs), this choice is also "wasteful." Then we can save a lot of computer time if we can parameterize W

directly. (This may involve several different sets of coordinates. Think of the unit circle in the plane. We can save bookkeeping space by labelling its points with a single polar coordinate θ instead of two rectangular coordinates x and y. However, we have to "change coordinates" from time to time so that a single point is not simultaneously labelled "0" and "2π".)

In practice, efficiently parameterizing W involves understanding its geometry in a very deep way. (This is the same problem addressed by Brockett, in a different context, in [Br].) This problem would seem almost hopeless, were it not for the fact that we know something about the map ω. In almost all interesting cases, ω is a rational map--that is, it is given by a set of rational functions in the coordinates. We know something about this rational map--it is the solution to a complicated problem in dynamic programming. Of course, we will never really understand the mapping until we understand the problem that it solves, but even its more obvious properties can be useful in understanding the geometry of its image.

* * *

As we have noted, the questions of when identification has been achieved come down to questions about the injectivity of ω and τ. Because a generic (or "almost everywhere") answer would satisfy us, and because these maps are rational, we are essentially asking whether the maps are what a geometer would call "birational." Here again geometric criteria can be applied--although their value-added is in doubt given the usefulness of the sorts of criteria described in footnote 7.

The geometry becomes crucial, however, in the case of problems which are known *not* to be identified. A simple example is the undiscounted optimal linear regulator problem: to find a feedback rule that will maximize

$$E\left[\left(\sum_{t=1}^{M-1} R(x_t) + Q(u_t)\right) + P(x_m)\right]$$

with R, Q and P quadratic and negative definite, subject to a linear control law

$$x_{t+1} = Ax_t + Bu_t + z_t$$

where z_t is white noise.

By standard techniques, the optimal control law is found to be

$$u_t = Fx_t$$

where $F = -(Q + B'PB)^{-1}B'PA$.

If agents adopt this law, the observable behavior of the economy will be

$$x_{t+1} = (A + BF)x_t + z_{t+1} \quad .$$

Given the observed behavior $A + BF$ (thought of as a point in T), we wish to calculate the working economy consisting of A, B and F and the ecomony consisting of A, B, P, Q and R. Dimensionality considerations suffice to reject this possibility out of hand. There are many possible choices of $(A,B,P,Q,R) = e$.

However, we can ask what kind of identifying restrictions we need in order to "naturally" associate to each $t \in T$ a unique $e \in E$. If we write ϕ for the composite map $\tau\omega : E \to T$ that associates to each economy its observable behavior, then what we are seeking is a map $\psi : T \to E$ with the property that $\psi\phi$ is the identity map. In the language of the geometer, we wish to construct a *section* for ϕ. In the language of the systems theorist, we seek a *canonical form* for e.

To the best of my (possibly archaic) knowledge, next to nothing is known about this problem. The choice of canonical form should surely be guided by economic considerations. But the questions of existence and explicit construction are geometric questions, and they are questions of the sort which a lot of geometric machinery has been constructed to tackle.

* * *

Geometry also has a lot to say about just how important the choice of canonical form is for policy evaluation. Consider again the map $\phi : E \to T$. The rational expectations recipe for policy evaluation is this: (1) Work in a context in which ϕ is one-to-one. Call its inverse ψ. (2) If z_0 is the policy to be evaluated, construct a map $\rho = \rho_{z_0} : E \to E$ by sending (v,s,z) to (v,s,z_0). (3) If t is the current description of the observed behavior of the economy, then the observed behavior under a policy of z_0 will be $\phi(\rho(\psi(t)))$.

The policy evaluation consists of evaluating various quantities $\xi(\phi(\rho(\psi(t))))$, where ξ is a "social welfare" function or some other function that the social planner is trying to maximize.

In the case in which full identification has not been achieved (ϕ not one-to-one) we can still hope to say something non-trivial about policy. Choosing a particular canonical form ψ may involve carrying more ideological baggage than we wish to burden ourselves with. (Depending on the context, it also may not.) But some statements about the effects of policy should be independent of the choice of ψ. In other words, we can look for opportunities to make statements like "I do not know what agents are maximizing. But in light of observed behavior, I can narrow it down to a small set of possibilities. This information is sufficient for me to conclude that policy z_0 is a very bad (or a very good) idea."

All of this is to say that we are looking for properties of $\xi(\phi(\rho(e)))$--such as, ideally, its value--which are invariant as e ranges not over the set of all economies but just over the small set of economies satisfying $\phi(e) = t$. We are asking for what functions ξ we can make such statements.

Now this again is the sort of problem that geometers have thought very hard about for very many years. In part, we are dealing with theory of invariants, the material of Hilbert's fourteenth problem.

For purposes of policy evaluation, one really wants to know the *distribution* of $\xi(\phi(\rho(e)))$ as e ranges over the fiber $\phi^{-1}(t)$. Perhaps we are seeking a probabilistic invariant theory which has yet to be worked out.

* * *

Now, all of these remarks are very vague--necessarily so because my thoughts on these matters are very vague (although my thoughts are not as vague as these remarks are). But I think it is reasonably clear that all of these problems are really geometric problems and that it will therefore pay to use some real geometry in thinking about them. This is particularly true in the context of rational expectations, because the solutions to the dynamic programming problems involved induce mappings which are rational (and hence geometrically tractable) but highly non-linear (and hence geometrically non-trivial).

(There is one very different way in which geometry can be expected to enter econometrics: if the laws of motion, etc. are specified over a ring of convolution operators, then the geometry of the spectrum of that ring can be expected to play a major role. This is primarily of interest in continuous time.)

To anyone intrigued by these possibilities, I close with a warning: One thing that we have learned from the history of business cycle models is that technically competent econometrics insufficiently motivated by economic theory is, for many purposes, bad econometrics. Let the calculator beware.

FOOTNOTES

[1] Mitchell divided each "reference cycle" into nine stages. Stages one and nine were three-month periods centered around successive troughs, and stage five was a three-month period centered around the intervening peak. Stages two, three and four were periods of equal length inserted between period one and five; stages six, seven and eight were defined similarly. Various time series are then found to follow cycles which are displaced from the reference cycle by a number of stages which is constant cycles (e.g. bank clearings expand from stages 8 and 4 to contract from stages 4 to 8; commercial paper rates expand from stages 2 to 6 and contract from stages 6 to 2--etc.). The "recurrent character of the business cycle" refers both to this observation and to certain persistent regularities in the amplitudes of the various cycles.

[2] Thus, in order to maintain high levels of employment, the monetary authority must consistently maintain a prime level that is higher than expected. As people come to expect this policy, it is hard to imagine any kind of an equilibrium.

[3] Absolutely the best non-technical discussion of modern equilibrium business cycle theory is [L1]. It is required reading for anyone who wishes to really understand the issues I have touched on in this section.

[4] It might be objected that this difficulty can be overcome by including an exogenous "policy variable" on the right-hand side of the regression. I am inclined to believe that there are cases in which this simple expedient will suffice, but it is no panacea. If the problem is to evaluate the consequences of a departure from what has been a stable policy, then there will be no data capable of distinguishing the new variable's coefficient. (To my knowledge all significant empirical work in the "rational expectations" literature assumes that policy has been stable throughout the sampling period. It may prove true that "policy variables" can be made to work when policy has been sufficiently variable, and that more radical remedies--such as are discussed in the remainder of this paper--are needed otherwise.) There are at least two other technical problems with this approach. One is that policies are often not *observed* by the econometrician, but rather *estimated* by him. (Think of the "average rate of inflation," for example, as a policy variable.) Thus this approach requires that the *coefficients* of one equation appear as *variables* in another--certainly a departure from the usual regression model, but presumably still within the purview of techniques like maximum likelihood. The other problem may be more fundamental: it is not clear that government policy is exogenous. In fact, government reacts to the behavior of private agents, and a change in policy sometimes consists of a change in the pattern of those reactions. In that case, policy must be counted as endogenous, and its inclusion only make identification more difficult.

[5] One of my colleagues in the mathematics department was amused by the introduction to [S], which describes the purpose of the paper as exploring "some of the implications for econometric practice of the principle that people's observed behavior will change when their constraints change." I had to explain to him that when this principle first entered econometric practice in the last decade, it was a radical departure from tradition!

[6] The right way to do all of this is to have people maximize given a perceived distribution of values for π. I have assumed that the distribution is always concentrated at a point, just to keep things simple.

[7] This is usually accomplished in practice by assuming that the process z is of large dimension and that it contains a lot of information which only enters the laws of motion after considerable lags.

[8] It is sometimes taken as an assumption that the space H being maximized over consists only of linear functions.

REFERENCES

[B] Robert J. Barro, *Unanticipated money growth and unemployment in the United States*, American Economic Review 67 (January, 1977), 101-115.

[Br] Roger Brockett, *Some geometric questions in the theory of linear systems*, IEEE Transactions in Automatic Control, AC-21 (1976).

[By] Christopher Byrnes, *The moduli space for linear dynamical systems*, in Geometric Control Theory (C. Martin and R. Hermann, eds.).

[F] Franklin Fisher, *The Identification Problem in Econometrics*, McGraw-Hill, 1966.

[Fr] Milton Friedman, *The role of monetary policy*, American Economic Review 58 (March 1968), 1-17.

[HS] Lars P. Hansen and Thomas J. Sargent, *Formulating and estimating dynamic linear rational expectations models*, Journal of Economic Dynamics and Control 2 (January, 1980) 7-46.

[L] Robert E. Lucas, Jr., *Studies in Business Cycle Theory*, MIT Press, 1981.

[L1] Robert E. Lucas, Jr., *Understanding business cycles*, in Stabilization of the Domestic and International Economy, vol. 5 of Carnegie-Rochester Series on Public Policy, North Holland, 1977. (Reprinted in [L]).

[L2] Robert E. Lucas, Jr., *Methods and problems in business cycle theory*, Journal of Money Credit and Banking 12 (November, 1980). (Reprinted in [L]).

[L3] Robert E. Lucas, Jr., *Econometric policy evaluation: a critique*, in The Phillips Curve and Labor Markets, vol. 1 of Carnegie-Rochester Series on Public Policy, North-Holland, 1976. (Reprinted in [L]).

[L4] Robert E. Lucas, Jr., *Expectations and the neutrality of money*, Journal of Economic Theory 4 (April, 1972) 103-124. (Reprinted in [L]).

[L5] Robert E. Lucas, Jr., *An equilibrium model of the business cycle*, Journal of Political Economy 83 (December, 1979) 1113-1144. (Reprinted in [L]).

[La] Steven E. Landsburg, *Real effects of price level variability*, to appear.

[M] Wesley Mitchell, *What happens during business cycles*, National Bureau of Economic Research, 1951.

[P] Edmund S. Phelps, *Money wage dynamics and labor market equilibrium*, Journal of Political Economy 76 (July, 1968) 687-711.

[S] Thomas J. Sargent, *Interpreting economic time series*, Journal of Political Economy 89 (April, 1981) 213-248.

Although it is not directly referenced in this paper, the following is an excellent source:

[LS] Robert E. Lucas, Jr. and Thomas J. Sargent, *After Keynesian macroeconomics*, in After the Phillips Curve: The Persistence of High Inflation and High Unemployment, Federal Reserve Bank of Boston, 1978.

DEPARTMENT OF ECONOMICS
UNIVERSITY OF CHICAGO
CHICAGO, ILLINOIS
U.S.A.

ON-LINE STRUCTURE SELECTION FOR MULTIVARIABLE STATE SPACE MODELS

A. J. M. van Overbeek and Lennart Ljung

ABSTRACT. An algorithm is described for the selection of model structure for identifying state space models. The algorithm receives as "input" a given system in a given parametrization. It is then tested whether this parametrization is suitable (well conditioned) for identification purposes. If not, a better one is selected and the transformation of the system to the new representation is performed.

This algorithm can be used as a block both in an iterative, off-line identification procedure, and for recursive, on-line identification. It can be called whenever there is some indication that the model structure is ill-conditioned. It is discussed how the model structure selection algorithm can be interfaced with an off-line identification procedure. A complete procedure is described and tested on real and simulated data.

1. INTRODUCTION. The procedure of identifying a system with several outputs contains some non-trivial problems regarding the parametrization of the model. Partly because of these, multivariable systems identification has not yet become a standard task for the user.

Several approaches to the parametrization or model structure selection problem have been taken. In this paper, we shall study the use of overlapping, parameter identifiable model structures. Such were discussed e.g. in Rissanen and Ljung [1975] and Ljung and Rissanen [1976].

With overlapping model structures it is possible to represent a given system within different structures. The question of which one to choose can be answered on the basis of information about the system. As this information improves, either in an iterative, off-line procedure or in a recursive, on-line algorithm, the decision about the best model structure may be revised and a change of model structure has to be performed.

We shall in this paper concentrate on the problem when and how to change between overlapping parametrizations. The idea is to create a procedure that is attached to the identification algorithm. This procedure monitors the conditioning of the parametrization. If the current structure is found to be bad, it decides which new one to switch to and how to calculate the representation within the new structure.

The paper is organized as follows. A brief discussion of approaches to the parametrization problem is given in Section 2. The particular approach that is adopted in this paper is described in more detail in Section 3. It is based on overlapping parametrizations. Section 4 is devoted to a discussion of how to detect and diagnose bad conditioning. The main result is that the properties of the parametrization can be inspected in terms of the state covariance matrix. This allows for a substantial saving in computing effort compared to analysis of the higher-dimensional second derivative matrix of the criterion. In Section 5 a procedure that monitors and selects the "best" conditioned parametrization is presented. It is based on the ideas of Section 3 and 4. The interfacing of this procedure to the identification algorithm is discussed in Section 6. In that section a complete algorithm for multivariable, black box identification is described. Applications to simulated and real data are also given.

2. PARAMETRIZATION OF MULTIVARIABLE SYSTEMS. Typical representations of multivariable stochastic, dynamical systems are as vector difference equations

$$y(t) + A_1 y(t-1) + \dots + A_{m_a} y(t-m_a) = B_1 u(t-1) + \dots + B_{m_b} u(t-m_b)$$
$$+ e(t) + C_1 e(t-1) + \dots + C_{m_c} e(t-m_c) \quad (2.1)$$

or state space models

$$\begin{aligned} x(t+1) &= Fx(t) + Gu(t) + Ke(t) \\ y(t) &= Hx(t) + e(t) \quad . \end{aligned} \quad (2.2)$$

Here $y(t)$ is the p-dimensional output of time t (a column vector), $u(t)$ is the input, and $e(t)$ is a sequence of independent random, p-dimensional vectors, with zero mean values and covariance matrix Λ. $x(t)$ is the n-dimensional state vector at time t, and A_i, B_i, C_i, F, G, K and H are matrices of compatible dimensions.

The representations (2.1) and (2.2) are equivalent in the sense that any system that can be represented in the form (2.1) can also be given as (2.2) with a certain n, and vice versa. There is however no direct relationship between the numbers n and m_a, m_b, m_c. The computational complexity of (2.1) when using it, e.g. for simulation, in terms of required memory space, best organization or calculations, etc., is actually given by the order of the corresponding state space model (2.2). Therefore, we shall in the sequel concentrate on (2.2).

In most of the treatment here, we shall assume that no input is present, $G = 0$. Presence of an input does not add any conceptual difficulties to the parametrization problem. It may however have an important influence on which parametrization is the best conditioned one. We shall make some comments on this matter in Section 6, after having tested the procedure on real data with an input signal.

The objective of the identification procedure is to determine n and the matrices F, K and H in (2.2) from measured data $\{y(t)\}$ so that an acceptable description of the data is obtained. The usual way to do this, is to parametrize the matrices, for given n, by a finite dimensional parameter vector Θ,

$$\begin{aligned} x(t+1) &= F(\Theta)x(t) + K(\Theta)e(t) \\ y(t) &= H(\Theta)x(t) + e(t) \\ &\Theta \in D_M \end{aligned} \tag{2.3}$$

and then minimize the prediction error loss function

$$V_N(\Theta) = h\left(\frac{1}{N}\sum_{t=1}^{N} \ell(t,\Theta,\varepsilon(t,\Theta))\right) \; ; \; \Theta \in D_M \tag{2.4}$$

where the prediction error is

$$\varepsilon(t,\Theta) = y(t) - \hat{y}(t|\Theta) \quad .$$

Here the prediction $y(t|\Theta)$ is computed from previous values of y as

$$\begin{aligned} \hat{x}(t+1,\Theta) &= F(\Theta)\hat{x}(t,\Theta) + K(\Theta)(y(t) - H(\Theta)\hat{x}(t,\Theta)) \\ \hat{y}(t,\Theta) &= H(\Theta)\hat{x}(t,\Theta) \quad . \end{aligned} \tag{2.5}$$

Normally we shall choose

$$\ell(t,\Theta,\varepsilon) = \varepsilon\,\varepsilon^T \qquad \text{(a } p|p\text{-matrix)} \tag{2.6a}$$

and

$$h(A) = \det A \quad . \tag{2.6b}$$

From the parametrization or model structure (2.3) we require two things: That there should be a Θ^* such that $V_N(\Theta)$ is minimized and that this minimizing value is unique, at least locally. The latter property is desirable for numerical reasons when $V_N(\Theta)$ is minimized. (Normally second order methods are used, and if the Hessian (the second derivative matrix) is singular at the minimum, numerical problems arise.)

There are, of course, many model structures that meet these two requirements. The problem is that some a priori information about the system is needed in order to select one such structure. For single output systems only the order of the system has to be known for this purpose. This does not pose a practical problem, since one may always test a sequence of models with increasing order.

For multi-order systems the structure is more detailed, and for each order there are many model structures that would have to be tried.

To solve this problem, some different approaches have been suggested in the literature. We may distinguish the following ones.

1. Physical a priori knowledge about the system is used to determine a parametrization (2.3). This approach can be taken when the system dynamics is known up to certain physical constants. Then Θ corresponds to these constants and the state vector $x(t)$ has a direct physical interpretation. This situation is no doubt the most favourable one. Physical insight will keep the number of parameters to be estimated small as well as give a concrete solution to the structure selection problem. Often, however, the system has to be treated as a black box, and then this approach is inapplicable.
2. Another approach is to use some very special parametrization that are guaranteed to be uniquely identifiable as soon as the state dimension is chosen appropriately. For state space models (2.2), this typically means that the F matrix is chosen to be diagonal, while for the model (2.1), one may restrict the A or C matrices to be diagonal (cf. Kashyap and Rao [1976]). With this approach, the structure selection problem is of the same type as for single-output systems. A disadvantage is that the restrictions imposed usually necessitates high dimensions of x and Θ. The method may, therefore, not be very efficient.
3. The approach that has attracted most interest is to consider various "canonical parametrizations." It is well known (see e.g. Luenberger [1967] or Rissanen [1974]) that some structure indices (like the observability or Kronecker indices) can be defined based on properties of the impulse response of the system. (A few details on this will be given in the next section.) With these indices, in turn, it is possible to define parametrizations within which the system can be uniquely represented. Therefore, if the structure indices are known, minimal state space realizations that are uniquely parametrized can be given to solve the problem we posed. There is a rather extensive literature on various aspects of this procedure. Typically, the structure indices are first estimated (by some rank testing procedure) and then the corresponding "canonical form" is chosen for identifying the system. (See, for example, Rissanen [1974], Akaike [1976], Guidorzi [1975] or Tse and Weinert [1975].) This approach gives a nice solution to the parametrization problem. Two points of criticism can be raised against it, however. First, rank testing to determine the structural properties is not a very well conditioned procedure, in particular if the signal-to-noise ratio is not too good. Second, this approach cannot be used in connection with recursive (on-line) identification since it requires a phase of a priori analysis of the data set.

4. With the "canonical form" representation described above, a given system can be represented within exactly one model structure and this representation is unique. From the identification point of view, the important issue lies in the uniqueness within the chosen representation. Whether other (unique) representations of the given system exist does not play any role. This point was stressed by Glover and Willems [1974]. The idea was pursued further in Rissanen and Ljung [1975] and Ljung and Rissanen [1976], whose different criteria for how to choose between different possible models structures also are discussed. The important feature of this point of view is that these parametrizations overlap, so that a change of parametrization can be made, without loss of information. This is described in more detail in the next section.

Based on these ideas, we may conceive a procedure for identifying both structure and parameters of a multi-output system: Start with any of the parametrizations as in (3.9), and minimize the prediction error criterion (2.4). Suppose that in the course of the minimization procedure it appears that the parametrization is ill-conditioned. Then a new one can be selected, and the corresponding similarity transformation can be computed, in order to determine the parameter values in the new parametrization. With these, the minimization of (2.4) can continue. Since the new system is equivalent to the previous one, the minimization effort spent in the first parametrization is not wasted. It may, of course, happen that a change of parametrization may take place a couple of times, before the minimum of (2.4) in a suitable structure is reached.

The crucial step in the above procedure is a program block which receives a particular model structure with given parameter values, tests if it can find a better model structure, if so, selects one and performs the transformation to the new structure. The output of the block is a new model structure with given parameter values, that is equivalent to the input one, but corresponds to a better conditioned basis.

The procedure can be depicted as in Figure 3.1.

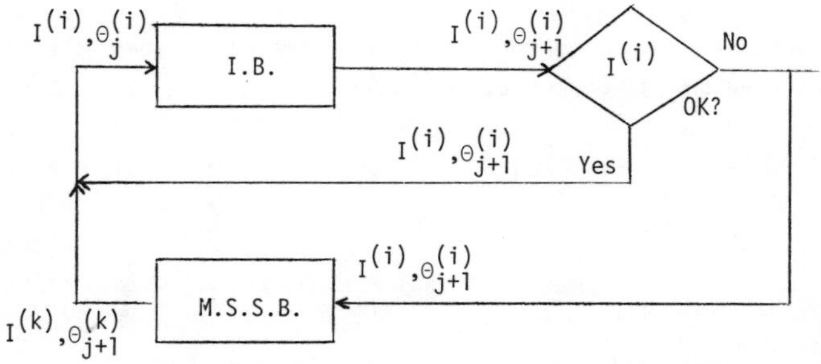

Fig. 3.1. On-line model structure selection for identification. I.B.: "Identification block, iteration or recursion nr. M.S.S.B.: Model structure selection block.

3. OVERLAPPING, IDENTIFIABLE PARAMETRIZATIONS. It is quite well known (see e.g. Kalman et al. [1969], Rissanen [1974] or Akaike [1974]) that the Hankel matrix of the impulse response can be used as a starting point for describing the system as a state space representation. For future reference, we shall briefly review the procedure here.

Let the impulse response (relating e to y) of the system

$$\begin{aligned} x(t+1) &= Fx(t) + Ke(t) \\ y(t) &= Hx(t) + e(t) \end{aligned} \tag{3.1}$$

be given by R_i $i = 0,1,\ldots,(R_0 = I)$. From this sequence, the blockwise Hankel matrix

$$R^r = \begin{pmatrix} R_1 & R_2 & \\ R_2 & R_3 & \cdots \\ \vdots & \vdots & \\ R_r & R_{r+1} & \cdots \end{pmatrix} \tag{3.2}$$

can be defined. Then, if the process $y(t)$ has an n-dimensional representation (3.1), all matrices R^r, $r \geq 1$ will have rank less or equal to n. The n first linearly independent rows determine the observability of Kronecker indices, and they can be used as a basis for representations like (3.1).

We have

$$y(t) = \sum_{k=0}^{\infty} R_k e(t-k)$$

and for $\hat{y}(t|t-r)$, the prediction of $y(t)$ based on $y(t-r)$, $y(t-r-1),\ldots,$

$$y(t|t-r) = \sum_{k=r}^{\infty} R_k e(t-k) \quad .$$

Hence

$$Y_N(t) \triangleq \begin{pmatrix} y(t|t-1) \\ \vdots \\ y(t+N|t-1) \end{pmatrix} = R^N \begin{pmatrix} e(t-1) \\ e(t-2) \\ \vdots \end{pmatrix} \tag{3.3}$$

Suppose now that rows $i_1,\ldots,i_n$ of R^N form a row basis for this matrix. We shall denote this index set by

$$I^{(i)} = \{i_1,\ldots,i_n\}$$

and use (i) as a superscript below, to emphasize that the quantities refer to this particular basis. Extract from $Y_N(t)$ the corresponding rows to form the column vector $x^{(i)}(t)$. Then, since $x^{(i)}(t)$ is a basis, all rows of $Y_N(t)$ can be expressed as linear combinations of $x^{(i)}(t)$

$$Y_N(t) = \mathcal{O}^{(i)} x^{(i)}(t) \quad . \tag{3.4}$$

In particular, the rows

$$i_1 + p, \ldots, i_n + p \quad \text{of} \quad Y_N(t)$$

can be expressed as $F^{(i)}x^{(i)}(t)$. The corresponding rows of $Y_N(t+1)$ are, of course $x^{(i)}(t+1)$, and since

$$\hat{y}(t+k|t) = \hat{y}(t+k|t-1) + \bar{K}_k e(t)$$

we find that

$$\begin{aligned} x^{(i)}(t+1) &= F^{(i)}x^{(i)}(t) + K^{(i)}e(t) \\ y(t) &= H^{(i)}x^{(i)}(t) + R_0 e(t) \quad . \end{aligned} \tag{3.5}$$

The row basis $I^{(i)}$ for R^N thus defines a unique parametrization for the state space representation. The parameters of $F^{(i)}$ are coordinates in the chosen basis, and they are therefore unique, as long as the rows in $I^{(i)}$ indeed are linearly independent.

It is also clear from (3.4) and (3.5) that $\mathcal{O}^{(i)}$ in fact is the observability matrix for (3.5):

$$\mathcal{O}^{(i)} = \begin{pmatrix} H^{(i)} \\ F^{(i)}H^{(i)} \\ \vdots \\ (F^{(i)})^{n-1}H^{(i)} \end{pmatrix} \tag{3.6}$$

In the sequel we shall confine ourselves to row bases I such that rows $1,\ldots,p$ belong to I and such that if row $k \in I$, then also row $k-p$ (if greater than zero) belongs to I.

We shall now illustrate the idea of overlapping parametrizations by a simple example.

EXAMPLE 3.1. Suppose the number of outputs (i.e., p) is 2, and the maximum rank of R^r (i.e., n) is 4. If we search from above in R^r for the 4 first linearly independent rows, the numbers of these rows may be

$$I^{(1)} = \{1,2,3,4\} \quad \text{or} \quad I^{(2)} = \{1,2,3,5\} \quad \text{or} \quad I^{(3)} = \{1,2,4,6\} \quad .$$

The corresponding parametrizations of the F-matrix are

$$F^{(1)} = \begin{pmatrix} 0 & 0 & 1 & 0 \\ 0 & 0 & 0 & 1 \\ x & x & x & x \\ x & x & x & x \end{pmatrix},$$

$$F^{(2)} = \begin{pmatrix} 0 & 0 & 1 & 0 \\ x & x & x & 0 \\ 0 & 0 & 0 & 1 \\ x & x & x & x \end{pmatrix} \tag{3.7}$$

$$F^{(3)} = \begin{pmatrix} x & x & 0 & 0 \\ 0 & 0 & 1 & 0 \\ 0 & 0 & 0 & 1 \\ x & x & x & x \end{pmatrix}$$

where "x" denotes the free parameters (to be estimated). The K and H matrices are

$$K = \begin{pmatrix} x & x \\ x & x \\ x & x \\ x & x \end{pmatrix} \qquad H = \begin{pmatrix} 1 & 0 & 0 & 0 \\ 0 & 1 & 0 & 0 \end{pmatrix} \tag{3.8}$$

in all three cases. The zero in the second row of $F^{(2)}$ comes from the fact that it is known that the 4-th row of the Hankel matrix is a linear combination of the three ones above. This follows from the way the basis is chosen. It should be noticed that a given system of order 4 with 2 outputs admits one and only one of the representations (3.7), (3.8).

However, if the sets $I^{(i)}$ were merely selected as row bases for R^r, without being obtained by a search for linear independnet rows from above, other parametrizations would be obtained. They are

$$F^{(1)} = \begin{pmatrix} 0 & 0 & 1 & 0 \\ 0 & 0 & 0 & 1 \\ x & x & x & x \\ x & x & x & x \end{pmatrix},$$

$$F^{(2)} = \begin{pmatrix} 0 & 0 & 1 & 0 \\ x & x & x & x \\ 0 & 0 & 0 & 1 \\ x & x & x & x \end{pmatrix} \tag{3.9}$$

$$F^{(3)} = \begin{pmatrix} x & x & x & x \\ 0 & 0 & 1 & 0 \\ 0 & 0 & 0 & 1 \\ x & x & x & x \end{pmatrix} \tag{3.9}$$

(continued)

with $K^{(i)}$ and H as before. If $I^{(i)}$ is a basis, then $F^{(i)}$, $K^{(i)}$ and H provides a unique parametrization for the system. However, since different row bases can be selected for R^r, a given system may be represented within several of the parametrizations (3.9). (See e.g. Ljung and Rissanen [1976] or Rissanen and Ljung [1975] for further details on this.)

REMARK. Structures for vector difference equations (2.1) can be determined from a row basis of (3.2) in a similar fashion. There is, however, a complication in that a basis that does not necessarily consist of the *first* linearly independent rows may lead to structures where a component $y_i(t)$ may depend on future values $y(t+k)$. We shall exemplify this in Section 6, when we discuss initial estimates obtained from models (2.1) with $C_i = 0$.

There are some important implications of the above facts. The first one is that there is a choice between different parametrizations. To obtain a well conditioned parametrization, it should correspond to a well conditioned row basis for R^r (cf. Ljung and Rissanen [1976], where complexity measures are used). The second implication is that since "most" (i.e., except those on hypersurfaces in the parameter space) systems can be represented in either of the forms (3.9), it is possible to change parametrization by a similarity transform without loosing any information.

A discussion of the model structure selection algorithm MSS will be given in section 5, while the whole procedure is described in more detail in Section 6. Before that, however, we shall study some basic facts for testing the conditioning of the parametrization.

4. THE CONDITIONING OF THE PARAMETRIZATION. The most important aspect of the parametrization from the identification point of view is that the criterion should be well conditioned for numerical minimization. This, in essence, means that V'', the second derivative matrix of the criterion with respect to the parameters, should be well conditioned at the minimum. Since the parameters of the F-matrix can be seen as the coordinates in the chosen row basis, a well conditioned parametrization corresponds to a well conditioned basis. The basis can be evaluated in terms of $P^{(i)}$, the stationary covariance matrix of the i-th row basis, i.e. $x^{(i)}(t)$. $P^{(i)}$ is the stationary solution of

$$P(k+1) = F^{(i)}P(k)F^{(i)T} + K^{(i)}K^{(i)T} \tag{4.1}$$

assuming that $e(t)$ in (3.5) is a white noise process with an identity covariance matrix. Note that $P^{(i)}$ only is a $n \times n$ matrix, while V" is $n_\Theta \times n_\Theta$.

Define the finite controllability matrix $S^{(i)}$ of (3.5) as

$$S^{(i)} = [K^{(i)}, F^{(i)}K^{(i)}, \ldots, F^{(i)n-1}K^{(i)}] \tag{4.2}$$

Also define:

$$\psi^{(i)}(t) = -H^{(i)} \frac{dx^{(i)}(t)}{d\Theta^{(i)}} = \frac{d}{d\Theta^{(i)}} \epsilon\ (t,\Theta) \tag{4.3a}$$

$$\Lambda = E\{e(t)e(t)^T\} \quad \text{and} \tag{4.3b}$$

$$W^{(i)} = E\{\psi^{(i)}(t)\Lambda^{-1}\psi^{(i)T}(t)\} \quad . \tag{4.3c}$$

Note that in (4.3) $F^{(i)} - K^{(i)}H^{(i)}$ needs to be stable for $W^{(i)}$ to exist. In this section $e(t)$ is considered to be a given stochastic process from which $y(t)$ is generated via the model $\Theta^{(i)}$ using (3.5). The expectation operator in (4.3b) and (4.3c) is to be taken with respect to $e(t)$. The conditioning of $P^{(i)}$ depends on how well the n rows by their smallest eigenvalues tending to zero. If W is nearly singular then there exists a vector ℓ, $||\ell|| = 1$, such that $||\ell^T S||$ is small. Note that $||S||$ is not small. In the same way if W is nearly singular then there exists an n_Θ vector λ, $||\lambda|| = 1$, such that $\lambda^T W \lambda$ is small. Now we can reformulate the theorem in the following way:

a) if the model is almost noncontrollable, i.e. there exists a small ε $\varepsilon \geq 0$ such that $||\ell^T S|| \leq \varepsilon$ when W is almost singular, i.e. there exists a λ such that $\lambda^T W \lambda \leq c\varepsilon$, where c is independent of $\Theta \in B(\Theta^+, r)$;

b) if W is almost singular, i.e. there exists a small ε, $\varepsilon \geq 0$ such that $\lambda^T W \lambda \leq \varepsilon$ then S is also almost singular, i.e. there exists an ℓ such that $||\ell^T S|| \leq c\varepsilon$, where c is independent of $\Theta \in B(\Theta^+, r)$;

c) in particular, S is singular if and only if W is singular.

We will prove this version of the theorem. In the proof ℓ and λ are normalized vectors with norm 1. A statement of the form "α small implies β small" will mean that existence of a small $\varepsilon \geq 0$ such that $||\alpha|| \leq \varepsilon$ implies $||\beta|| \leq c\varepsilon$, where c is independent of $\Theta \in B(\Theta^+, r)$.

Let

$$w(t) = \psi(t)^T \lambda \tag{4.4}$$

taken with (4.3c):

$$\lambda^T W \lambda = E w^T(t) \Lambda^{-1} w(t) \quad . \tag{4.5}$$

We will use the impulse response of the system relating $w(t)$ to $e(t)$. The state vector of this system consists of the concatenation of $z(t)$ and $x(t)$. $z(t)$ is an n vector defined as

$$z(t) = \frac{dx(t)^T}{d\Theta} \lambda \tag{4.6}$$

$S^{(i)}$ span the n dimension space. In fact if one would take the infinite controllability matrix $S^{\infty(i)}$ then $P^{(i)} = S^{\infty(i)}S^{\infty(i)T}$. Thus $P^{(i)}$ ill conditioned means that the n rows of both $S^{\infty(i)}$ and $S^{(i)}$ do not span the n dimensional space very well, i.e. the model (3.5) is "almost noncontrollable."

The main part of this section is devoted to a theorem concerning the relation between the conditioning of $W^{(i)}$ and $S^{(i)}$. The connection between $W^{(i)}$, computed from $e(t)$ and the "true system," and the second derivative matrix, computed during the minimization from $y(t)$ and the actual model, will be discussed after the theorem. In the remainder of this section we will drop the superscript (i) denoting the i-th basis.

The question now is how are the conditioning of W and of S related when Θ varies? We will show that if W is singular at a point Θ^+ then also S is singular and vice versa. Also, if W and S are singular at Θ^+ then W and S get singular at the same rate when Θ approaches Θ^+, i.e. the ratio of their condition numbers is bounded by some finite number when Θ varies in a neighbourhood around Θ^+. This means that one can inspect P to test for W being almost singular. Now we can formalize this discussion into the following theorem.

THEOREM 4.1. *Let the model* Θ *be such that both* $F(\Theta)$ *and* $F(\Theta) - K(\Theta)H$ *are stable and let* $S(\Theta)$ *and* $W(\Theta)$ *be as defined in (4.2) and (4.3) respectively, then* $W(\Theta)$ *is singular at a point* Θ^+ *if and only if* $S(\Theta^+)$ *is singular. Moreover, in a neighbourhood* $B(\Theta^+, r)$ *around such a singularity point*

$$0 < \underline{C} \le \frac{k(W(\Theta))}{k(S(\Theta))} \le \bar{C} < \infty$$

where k *denotes the condition number of the matrix.*

<u>Proof</u>. In the neighbourhood $B(\Theta^+, r)$ the elements of W and S are bounded. A large condition number is therefore caused. Differentiation of (3.1) and using that H does not contain any elements of Θ gives:

$$z(t+1) = [F-KH]z(t) + L_1x(t) + L_2e(t) \tag{4.7a}$$

$$x(t+1) = Fx(t) + Ke(t) \tag{4.7b}$$

$$w(t) = Hz(t) \quad . \tag{4.7c}$$

L_1 is an $n \times n$ matrix containing np nonzero elements of λ on the same places as and corresponding to the free parameters of F. If f_{ij} is a free parameter of F then L_{1ij} is the element of λ that is multiplied with $-H\frac{dx(t)}{df_{ij}}$ in (4.4). L_2 is a full $n \times p$ matrix containing the elements of λ corresponding to and on the same places as the elements of the K matrix.

We will need the following lemma:

LEMMA 6.1. *Let* h(t) *denote the impulse response from* e(t) *to* w(t) *of (4.7). Then* $\lambda^T W\lambda$ *small implies* h(t) *small for all* t *and vice versa.*

Proof. Recalling (4.5) we get

$$\lambda^T W\lambda = Ew^T(t)\Lambda^{-1}w(t) \le \varepsilon \quad . \tag{4.8}$$

Further

$$w(t) = \sum_{s=-\infty}^{t} h(t-s)e(s) \quad \text{and} \quad H(e^{i\omega}) = \sum_{t=1}^{\infty} h(t)e^{i\omega t} \tag{4.9}$$

Since e(t) is a full rank white noise process, its spectrum $\Phi_e(\omega)$ is positive, i.e.

$$0 \le \underline{C}I < \Phi_e(o) \le \bar{C}I < \infty \tag{4.10}$$

Because Λ is positive definite $E\{w(t)^T w(t)\}$ is small and from this also $E\{w(t)w(t)^T\}$ is small (a positive semidefinite matrix with a small trace has all its elements small).

$$E\{w(t)w(t)^T\} = \int H(e^{i\omega})\Phi_e(\omega)H^T(e^{i\omega})d\omega \le \varepsilon \quad . \tag{4.11}$$

Now using (4.10) we get

$$\varepsilon/\bar{C} \le \int H(e^{i\omega})H^T(e^{i\omega})d\omega \le \varepsilon/\underline{C} \quad .$$

From this follows h(t) small. A similar argument can be used to show that h(t) small implies $\lambda^T W\lambda$ small. This completes the proof of the lemma.

The proof of a) is simple. Choose $L_2 = 0$ and put the p nonzero rows of L_1 equal to ℓ. Then z(t) in (4.7a) is driven by $L_1x(t)$. If e(t) is an impulse then $[x(1),x(2),\ldots,x(n)]$ is by definition the controllability matrix S. Since $\|L_1 s\| \le \varepsilon$, so is $\|L_1 x(t)\|$ for $t = 1,\ldots,n$. Because F is stable, $\|L_1 S(t)\|$ is small for all t. Since F - KH is stable, z(t) and w(t) are small. Since the input e(t) is an impulse w(t) = h(t) and thus h(t) is small for all t. Now using Lemma 6.1 gives $\lambda^T W\lambda$ small.

To prove b) we rewrite (4.7) as:

$$\begin{pmatrix} z(t+1) \\ \hline x(t+1) \end{pmatrix} = \left(\begin{array}{c|c} \Phi & L_1 \\ \hline 0 & F \end{array}\right) \begin{pmatrix} z(t) \\ \hline x(t) \end{pmatrix} + \begin{pmatrix} L_2 \\ \hline K \end{pmatrix} e(t)$$

$$w(t) = [H \quad 0] \begin{pmatrix} z(t) \\ \hline x(t) \end{pmatrix} \tag{4.12}$$

$$\Phi = F - KH$$

The impulse response can be written as $[H \quad 0] \; S^\infty$ where S^∞ is the infinite controllability matrix of (4.10):

$$S^\infty = \left(\begin{array}{c|c} \Gamma_1 & \Gamma_2 \\ \hline K & FK \end{array} \; \cdots\cdots \; \begin{array}{c} \Gamma_i \\ \hline F^{i-1}K \end{array} \; \cdots\cdots\right) \tag{4.13a}$$

with

$$\Gamma_1 = L_2, \quad \Gamma_{i+1} = \Phi\Gamma_i + L_1F^{i-1}K, \quad i = 1,2,\ldots \tag{4.13b}$$

Now we need the following lemma:

LEMMA 6.2. *The overlapping parametrizations described in section 3 are such that* $h(i) = H\Gamma_i$ *small for all* i *implies* Γ_i *small for all* i.

<u>Proof</u>. We have to look at the structure of H and F. Here we will use a different ordering of the basis vectors than in section 3. Group the basis vectors output by output, i.e., in example 3.1. $I^{(1)} = \{1,3,2,4\}$, $I^{(2)} = \{1,3,5,2\}$, $I^{(3)} = \{1,2,4,6\}$. These correspond to the pseudo observability indices $0^{(i)} = \{0_1,\ldots,0_p\}$, with $0^{(1)} = \{2,2\}$, $0^{(2)} = \{3,1\}$, $0^{(3)} = \{1,3\}$ respectively. The F and H matrices are:

$$F = \begin{pmatrix} F_{11} & \cdots & F_{1p} \\ \cdot & & \cdot \\ F_{p1} & \cdots & F_{pp} \end{pmatrix}$$

F_{ij} is an $0_i \times 0_j$ matrix

$$F_{ii} = \left(\begin{array}{c|c} 0 & I_{0_i - 1} \\ \hline f_{ii,1} & f_{ii,2} \cdots f_{ii} \end{array}\right) \qquad F_{ij} = \begin{pmatrix} 0_{0_i - 1, 0_j} \\ \hline f_{ij,1} \cdots f_{ij,0_j} \end{pmatrix} \tag{4.14}$$

$$H = \begin{pmatrix} H_{11} & \cdots & H_{1p} \\ \cdot & & \cdot \\ H_{p1} & \cdots & H_{pp} \end{pmatrix}$$

where H_{ij} is an 0_j row vector. $H_{ij} = 0$, $i \neq j$ $H_{ii}(j) = \begin{cases} 1 & j=0 \\ 0 & j>1 \end{cases}$.

Also define

$$s_i = \sum_{k=1}^{i} 0_k \ , \quad i = 1,\dots,p \quad s_0 = 0 \quad . \tag{4.15}$$

Then the impulse response consists of rows $s_{i-1} + 1$, $i = 1,\dots,p$ of S^∞ corresponding to the ones in H_{ii}. Rewrite (4.13b) as

$$\Gamma_{k+1} = F\Gamma_k + L_1F^{k-1}K - KH\Gamma_k \quad . \tag{4.16}$$

Define $\gamma_{j,k}$ as the j-th row of Γ_k. Suppose Γ_k contains a "large element" (large compared to ε) in row $\gamma_{j,k}$. We will show that this row will appear unchanged in $h(k+j-s_r-1)$, where s_r is the last s_i of (4.15) less than j, if the last term of the right hand side of (4.16) is dropped. Since this term is $Kh(k)$ it is small and it cannot change the "large element" very much until it appears in $h(k+j-s_r-1)$. Dropping $KH\Gamma_k$ in (4.16) gives:

$$\Gamma_{k+1} = F\Gamma_k + L_1F^{k-1}K \quad . \tag{4.17}$$

From the structure of F and L_1 it follows that only rows s_i, $i = 1,\dots,p$ of Γ_{k+1} contain new elements, the others are shifted rows from Γ_k. This shift property of (4.17) can be expressed as:

$$\gamma_{j+1,k} = \gamma_{j,k+1} \ , \quad s_{i-1} < j < s_i \ , \quad i = 1,\dots,p \quad . \tag{4.18}$$

The shift property (4.18) causes rows s_i of Γ_k to appear in the impulse response after $0_i - 1$ shifts, i.e., in $h(k+0_i-1)$.

This means that if the impulse response is small then also Γ_k is small if (4.17) is used. Using (4.16) does not change the argument because the extra term $KH\Gamma_k$ is small. This completes the proof of the lemma.

From Lemma 6.1 follows that (4.12) has a small impulse response, and from this and Lemma 6.2 that $\|\Gamma_i\|$ is small for all i. Then from (4.13b) $\|L_1F^{i-1}K\|$ small and from this $\|L_1S\|$ small. Taking one of the rows of L_1 as ℓ gives the desired result.

Part c) of the theorem follows directly by putting $\varepsilon = 0$ in the proof of parts a) and b).

What is the relationship between W and the second derivative matrix of the minimization criterion? In the theorem the generating process $e(t)$ was known. In an identification all quantities are generated from $y(t)$. We know that at the global minimum the residuals $\varepsilon(t,\Theta)$ have the same statistical properties as $e(t)$ if the "true system" is representable in the model set. During the minimization W can be computed by replacing $e(t)$ by the actual residuals and the expectation operator in (4.3) by the corresponding summation. Note that $\psi(t)$ in (4.3a) is $\frac{d\varepsilon(t)}{d\Theta}$. The matrix $\psi(t)\psi^T(t)$ obtained in this way is an approximation of the second derivative matrix that is used in the calculation of the search direction in a Gauss Newton type method, if one uses (2.4) as a criterion with $\ell(t,\Theta,\varepsilon) = \varepsilon^T(t)\varepsilon(t)$ and $h(\Lambda) = \Lambda$. Then $EV'' = E\psi(t)\psi^T(t)$ at the global minimum point if the "true system" is a member of the model set. If one uses the determinant criterion (2.4)-(2.6) then at the global minimum $EV'' = \det \Lambda \cdot W$.

The conditioning of W is closely connected with the numerical properties of the minimization algorithm. Therefore we may expect that testing the parametrization by inspecting the conditioning of P is a reasonable approach. How this can be utilized to select a better conditioned basis is described in the next section.

5. THE MODEL STRUCTURE SELECTION ALGORITHM. In order to select a better conditioned basis one could compute all possible bases and test the conditioning of each. Then one would select the best or at least a better conditioned one. Already for relative modest values of n and p this will lead to a substantial amount of computation. For example, for $n = 10$ and $p = 3$ there are 36 different parametrizations. Hence, we do not want to test all possible parametrizations. The only thing we want is a better conditioned basis to get out of detected or expected numerical problems. Therefore we use an algorithm based on the properties of the stationary state covariance matrix $P^{(i)}$ as described in the previous section.

$P^{(i)}$ is computed in L-D factorized form.

$$P^{(i)} = L^{(i)}D^{(i)}L^{(i)T} \quad \text{with } L^{(i)} \text{ lower triangular and} \tag{5.1}$$

$$D^{(i)} = \text{diag}\,\{d_1^{(i)}, d_2^{(i)}, \ldots, d_n^{(i)}\} \quad .$$

(5.1) is iterated to stationarity with the help of the acceleration formula

$$P(k+1) = A(k+1)P(k)A(k+1)^T + P(k) \tag{5.2}$$

with

$$A(k+1) = A(k)A(k), \quad A(0) = F^{(i)}, \quad P(0) = K^{(i)}K^{(i)}{}_T \quad .$$

In iteration (5.2) is updated in L-D factorized form. For details see Bierman [1977]. Convergence is obtained if both $||D(k+1) - D(k)||$ and $||A(k+1)||$ are small.

To test if the present parametrization is ill-conditioned, one can compute the complexity of $P^{(i)}$ or look at the diagonal elements of $D^{(i)}$. It is well known that if we orthogonalize the basis $x_1,\ldots,x_n$ of model structure i then the length of the orthogonal basis vectors is given by the diagonal elements of $D^{(i)} : d_j^{(i)}$. As a measure of the conditioning, e.g. the quotient between the largest and smallest element of $D^{(i)}$ can be used. A well conditioned P has the same order of magnitude for all elements of $D^{(i)}$. The test can also be done in the identification algorithm by monitoring e.g. the diagonal elements of the Hessian if one uses a second order minimization method with a square root type of algorithm. The computation of $P^{(i)}$ with (5.1)-(5.2) is the first step in the selection algorithm. We then want to test this matrix $P^{(i)}$ against the covariance matrix that would be obtained in another structure, corresponding to the index set, say $I^{(k)}$.

Let T_{ik} be the transformation matrix from parametrization i to parametrization k. Then $P^{(k)}$ is given by

$$P^{(k)} = T_{ik}P^{(i)}T_{ik}^T \quad . \tag{5.3}$$

The conditioning of $P^{(k)}$ can be computed by L-D factorizing $P^{(k)}$. If $P^{(i)}$ is given in L-D factorized form then

$$P^{(k)} = T_{ik}L^{(i)}D^{(i)}L^{(i)T}T_{ik}^T = L^{(k)}D^{(k)}L^{(k)T} \tag{5.4}$$

and $L^{(k)}$ and $D^{(k)}$ can be computed by orthogonalizing the rows of $T_{ik}L^{(i)}$ with a modified weighted Grahm-Schmidt (MWGS)-algorithm with weights $d_j^{(i)}$. See Bierman [1977].

$P^{(k)}$ is well conditioned if the lengths of the orthogonal basis vectors of $T_{ik}L^{(i)}$ (i.e. the $d_j^{(k)}$) are of the same order of magnitude. T_{ik}^j should have the property that the rows of $T_{ik}L^{(i)}$ define the n-dimensional space well.

To change to a new basis we must compute the new state variables (= predictors of a set of output components) in terms of the old ones. We will get all predictors of the output in structure i by premultiplying x with the observability matrix for structure i: $\mathcal{O}^{(i)}$. See (3.4). The transformation matrix thus consists of rows of the observability matrix. For example, let $n = 4$, $p = 3$, $I^{(1)} = \{1,2,3,4\}$ and $I^{(2)} = \{1,2,3,5\}$ then T_{12} consists of rows 1,2,3 and 5 of $\mathcal{O}^{(i)}$.

The basic idea of the selection algorithm is to select those rows of $W = \mathcal{O}^{(i)}L^{(i)}$ such that we get "good" $d_j^{(k)}$ when we orthogonalize these rows. T_{ik} then consists of the same rows of $\mathcal{O}^{(i)}$. The next problem is how to organize the search for rows of W. It is assumed that y is a full rank process. This means that the first p rows of W always are selected and all bases must be selected from the first $(n-p+1)p$ rows of W. Since our $I^{(i)}$ follow the same rules as for "canonical forms" (if row $kp+j$ is not selected: then rows $\ell p+j$ will not be selected for $\ell > k$) we organize the search in the following way.

Start with the first p rows. Because of the above described selection rule one of rows $p+1, p+2, \ldots, p+p$ must belong to the basis. Select the one which gives the largest element in $D^{(k)}$. Say it was row $p+2$. In the next step one of rows $p+1, 2p+2, p+3, \ldots, p+p$ must belong to the new basis. Test again for the largest $D^{(k)}$ element (i.e. the longest vector in the orthogonalizing procedure) etc. A typical selection procedure may look like this for $n=5$ and $p=2$.

Step	Selected previously	Candidates	Selected
1	1,2	3,4	3
2	1,2,3	5,4	5
3	1,2,3,5	7,4	4
4	1,2,3,4,5	---	---

In the algorithm we combine this selection with the MWGS-procedure by checking at each step the lengths of all p candidates and choosing the longest one. The details of the algorithm are given in the following:

1. Compute $P^{(i)} = L^{(i)}D^{(i)}L^{(i)T}$ by (5.2).
2. Compute the observability matrix $\mathcal{O}^{(i)}$ of model structure i. Dimension $p(n-p+1) \times n$, because all bases must lie in the first $n-p+1$ block rows since the first p rows (= the first block row of $\mathcal{O}^{(i)}$ = the first p rows of the $n \times n$ identity matrix) always belong to the transformation matrix.
3. Compute $W = \mathcal{O}^{(i)}L^{(i)}$.

The rows of this matrix are orthogonalized with a modified weighted Grahm-Schmidt algorithm with row selection.

4. No orthogonalization needs to be performed for the first p rows of W. Put $\ell = p$ and $o_j = 1$, $j = 1, \ldots, p$.
5. Put $\ell = \ell + 1$. Select the longest row of rows on $o_j p + j$ $j = 1, \ldots, p$ of W. The measure of length is the weighted norm with weights

   $$d_j^{(i)} \quad j = 1, \ldots, n.$$

   Let row $r = o_q p + j$ be the longest one.
6. Put $o_q = o_q + 1$ and exchange rows r and ℓ.
7. Subtract base vector component ℓ from the rows below row ℓ.
8. Go to 5 unless $\ell = n$.

REMARK 1. The row numbers $o_j p + j$ of point 5 of the algorithm are the row numbers in the original W-matrix. One has to do some bookkeeping in a permutation array to keep track of the original rows after a number of exchanges in point 6.

REMARK 2. As a result of the orthogonalization procedure we obtain a L-D factorized matrix. This can be the new covariance matrix if the rows of W are selected in the right order. For example, if the new basis has

$$I^{(k)} = \{1,2,3,4,5\} \qquad (n = 5,\ p = 3)$$

and the rows are selected in this order from W we will have the new $P^{(k)}$ in L-D factorized form. If they are selected in the order 1,2,3,5,4 we will not have $P^{(k)}$ but a matrix with the same conditioning as $P^{(k)}$. Note that this matrix is obtained without performing the actual similarity transformation.

REMARK 3. We do not claim that this algorithm selects the best structure of all possible ones. It only selects a better one without the need of testing all possibilities. A better conditioned structure is sufficient to continue the minimization of (2.4).

REMARK 4. The algorithm only works when $F^{(i)}$ is stable. If $F^{(i)}$ is not stable $P^{(i)}$ cannot be computed. One could modify the algorithm in the following way. Let $S^{(i)}$ be the controllability matrix of structure i. Compute $W^{(i)} = \mathcal{O}^{(i)} S^{(i)}$ and orthogonalize this matrix with the MWGS algorithm with row selection in the same way as in the original algorithm. In this way n rows are selected from the leftmost $n - p + 1$ block columns of the Hankel matrix R^{n-p+1} of (3.2).

The algorithm has been coded in FORTRAN IV on a DECsystem-10 computer. In the computation of the inner products in the MWGS algorithm the multiplications are done in single and the additions in double precision. Some results are given in the following example.

EXAMPLE 5.1. In Tse and Weinert [1975] the following innovations representation is used:

$$F' = \begin{pmatrix} 0 & 1 & 0 & 0 \\ -0.5 & 1.0 & 0 & 0 \\ 0 & 0 & 0 & 1 \\ -1.0 & -4.0 & -0.25 & 0.0 \end{pmatrix}$$

$$K' = \begin{pmatrix} -0.0398 & -0.312 & -0.0112 \\ -0.115 & -0.304 & -0.0399 \\ -0.798 & -0.236 & -0.215 \\ -0.00459 & 1.27 & -0.215 \end{pmatrix}$$

$$H' = \begin{pmatrix} 1 & 0 & 0 & 0 \\ -1.0 & -3.0 & 0 & 0 \\ 0 & 0 & 1 & 0 \end{pmatrix}$$

Since $n = 4$ and $p = 3$ the possible model structure are: $I^{(1)} = \{1,2,3,4\}$, $I^{(2)} = \{1,2,3,5\}$ and $I^{(3)} = \{1,2,3,6\}$. This system can only be represented in $I^{(3)}$ since the transformation matrices to both other structures are singular. The transformation matrix to $I^{(3)}$ is:

$$\begin{pmatrix} 1 & 0 & 0 & 0 \\ -1 & -3 & 0 & 0 \\ 0 & 0 & 1 & 0 \\ 0 & 0 & 0 & 1 \end{pmatrix} .$$

This gives

$$F^{(3)} = \begin{pmatrix} -0.333 & -0.333 & 0.0 & e \\ 2.833 & 1.333 & 0.0 & e \\ 0 & 0 & 0 & 1 \\ 0.333 & 1.333 & -0.25 & 0.0 \end{pmatrix} \quad \text{with} \quad e = 0$$

$$K^{(3)} = \begin{pmatrix} -0.0398 & -0.312 & -0.0112 \\ 0.3848 & 1.224 & 0.1309 \\ -0.798 & -0.236 & -0.215 \\ -0.00459 & 1.27 & -0.215 \end{pmatrix}$$

$$H = \begin{pmatrix} 1 & 0 & 0 & 0 \\ 0 & 1 & 0 & 0 \\ 0 & 0 & 1 & 0 \end{pmatrix}$$

To make the system representable in $I^{(1)}$ and $I^{(2)}$ we put $e = 10^{-5}$. This makes $I^{(1)}$ and $I^{(2)}$ possible, but ill conditioned structures. The model structures $I^{(1)}$ and $I^{(2)}$ then are:

$$F^{(1)} = \begin{pmatrix} 0 & 0 & 0 & 1 \\ 3.167 & 1.667 & 0.0 & 1.0 \\ 0.333\ 10^5 & 0.333\ 10^5 & 0.0 & 1.0\ 10^5 \\ -1.056 & -0.5555 & -0.25\ 10^{-5} & -0.6667 \end{pmatrix}$$

$$K^{(1)} = \begin{pmatrix} -0.0398 & -0.312 & -0.0112 \\ 0.3848 & 1.224 & 0.1309 \\ -0.798 & -0.236 & -0.215 \\ -0.115 & -0.304 & -0.040 \end{pmatrix}$$

$$F^{(2)} = \begin{pmatrix} -3.167 & -1.667 & 0.0 & 1.0 \\ 0 & 0 & 0 & 1 \\ -2.833\ 10^5 & -1.333\ 10^5 & 0.0 & 1.0\ 10^5 \\ -8.972 & -4.722 & -0.25\ 10^{-5} & 4.107 \end{pmatrix}$$

$$K^{(2)} = \begin{pmatrix} -0.0398 & -0.312 & -0.0112 \\ 0.3848 & 1.224 & 0.1309 \\ -0.798 & -0.236 & -0.215 \\ -0.4003 & -0.748 & 0.1428 \end{pmatrix}$$

The results when these structures are used as input to the algorithm are given in Table 5.1.

The complexity is a measure for the conditioning of a matrix. The complexity of P is defined as $-\sum_{k=1}^{n} \ln(n\lambda_k)$, where λ_k are the eigenvalues of $P/\mathrm{trace}(P)$.

The algorithm selects the correct structure in all three cases. One expects the upper left 3×3 corner of L and D to be the same for all structures since $p = 3$. The differences are caused by the convergence criterion for the solution of (5.2) and roundoff errors. In spite of the bad conditioning of $I^{(1)}$ and $I^{(2)}$, indicated by the large values of the complexity, the system matrices after the transformation are very well in agreement with $F^{(3)}$ and $K^{(3)}$.

In the next section we will combine the model structure selection algorithm with an identification algorithm and apply it on both simulated and real data.

6. COMBINING THE MODEL STRUCTURE SELECTION ALGORITHM WITH AN IDENTIFICATION ALGORITHM. There are some questions which must be answered when the model structure selection (MSS) algorithm is combined with an identification algorithm. This section will deal with a few of these and it contains two examples showing the functioning of the combined algorithm when applied to both simulated and real data.

How often should the model structure selection algorithm be called, i.e. when is the question "$I^{(i)}$ OK?" in Figure 3.1 answered with no? If an off-line method is used one can call the MSS algorithm after each iteration, since the computing time will be dominated by the computation of the loss function, the gradient and depending on the method, the second derivative matrix. If a recursive method is used where the data is processed one data point at a time, one should monitor the second derivative matrix or its approximation and only call the MSS algorithm when this measure indicates a bad conditioned parametrization. Also one could ask if some kind of hysteresis must be included in the MSS algorithm to avoid too many parametrization changes.

When a parametrization change takes place then not only the parameters but also the gradient and the second derivative matrix should be changed or recomputed using the new parameters. If the gradient and (an approximation to) V" are computed in each iteration (i.e. in a Gauss-Newton type algorithm) this is no problem. However, if one uses a quasi-Newton type algorithm, which builds an approximation of V" over several iterations, one can save computing time by also changing V" instead of the brute force method of resetting the Hessian to the unit matrix. How this can be done is outlined below.

The new parametrization is obtained from the old one by means of a similarity transformation T_{ik}. The new parameters are not a linear function of the old ones, since T_{ik} itself is a function of the old parameters. Denote the old parameters by a and the new ones by b then $b = b(a)$ and $a = a(b)$.

$$\frac{dV}{db_i} = \frac{dV^T}{da} \frac{da^T}{db_i} \tag{6.1}$$

and

$$\frac{d^2V}{db_i db_j} = \frac{d}{db_j} \left\{ \frac{dV^T}{da} \frac{da^T}{db_i} \right\} = \frac{d^2V^T}{db_j da} \frac{da^T}{db_i} + \frac{dV^T}{da} \frac{d^2a^T}{db_i db_j} \tag{6.2}$$

The last term in (6.2) is zero if a is a linear function of b, i.e. if $a = Ub^T$. The first term can then be written as

$$\frac{d^2V}{db_i db_j} = \frac{da}{db_j} V'' \frac{da^T}{db_i} = U^T V'' U \tag{6.3}$$

If we just take the linear terms in $a = a(b)$, i.e. if we assume that the transformation matrix is not a function of the parameters, then the Hessian can be changed by the following procedure. Let $H^{(i)}$ denote the old Hessian, $H^{(k)}$ the new one and let $H^{(i)}$ be given in L-D factorized form, i.e. $H^{(i)} = L^{(i)}D^{(i)}L^{(i)T}$. First compute the matrix U from T_{ik} and T_{ki}, then compute $W = U^T L^{(i)}$ and triangularize W again with the MWGS algorithm using $D^{(i)}$ as the weights. It can be shown that then $U^T L^{(i)} D^{(i)} L^{(i)T} U = L^{(k)} D^{(k)} L^{(k)T} = H^{(k)}$. Compare with (3.5).

If the parameter vector consists of the p parameter rows of F followed by the p columns of K then the U matrix is block diagonal: $U = \operatorname{diag}\{UF, UK_1, \ldots, UK_p\}$. UF is a $np \times np$ matrix and a function of both T_{ik} and T_{ki}. The UK_i matrices are $n \times n$ and identical with T_{ki}. This procedure should function better in terms of saving iterations than just putting $H^{(k)}$ equal to the identity matrix. Taking also the nonlinear terms of $a = a(b)$ makes the computation far more complicated and the question is if it is worth the price.

The identification method of the combined algorithm used in the examples below is a quasi-Newton method. The loss function is the determinant criterion (2.4)-(2.6). The minimization routine is FORTRAN subroutine E04DDF from the NAG library which in turn is based on the Algol procedure QMNDER described in Gill and Murray [1972] and in Gill, Murray and Pitfield [1972]. The major modification made consists of a test for stability of $F - KH$ when a step is made in the line search routine. If the step makes the predictor unstable the step length is halved until $F - KH$ is stable. The parametrization used is (4.14). The MSS alogrithm is called after each iteration. The MSS algorithm only works if F is stable. If F is not stable the MSS algorithm is not called. Compare with Remark 4 in Section 5. When the parametrization is changed the loss function and the gradient are recomputed, since the algorithm requires exact derivatives and the derivative routine needs the states computed by the loss function routine. The approximative Hessian is reset to the unit matrix. In all identifications the minimization of the loss function was terminated when the gradient norm was less than 10^{-6} or when the quasi-Newton algorithm could not find a lower value of the loss function.

EXAMPLE 6.1. The identification is done on 1000 datapoints generated by the following system:

$$F = \begin{pmatrix} 0 & 0 & 1 \\ 0.25 & 0.0 & 0.25 \\ 1.0 & 0.0 & 0.0 \end{pmatrix} \qquad K = \begin{pmatrix} 0.547\ 10^{-2} & 0.063 \\ 0.119 & 0.157 \\ 0.674 & 0.666\ 10^{-3} \end{pmatrix}$$

$$H = \begin{pmatrix} 1 & 0 & 0 \\ 0 & 0 & 1 \end{pmatrix}$$

with a white noise input. The noise has an identity covariance matrix. The two possible model structures are $I^{(1)} = \{1,2,3\}$ and $I^{(2)} = \{1,2,4\}$ or $O^{(1)} = \{2,1\}$ and $O^{(2)} = \{1,2\}$ respectively. The system can only be represented in $O^{(1)}$. The results of the identification with a zero initial parameter estimate and when $O^{(1)}$ and $O^{(2)}$ are given as initial model structures are given in Tables 6.1 and 6.2. The results of an identification where the structure was fixed to $O^{(2)}$ are also included for comparison.

We see that the combined algorithm converges to the same point when the parametrization is allowed to change. The number of iterations in case $0^{(2)}$ could certainly be reduced if the Hessian was changed with the algorithm described earlier in this section. In case of the fixed parametrization no minimum was found within the given maximum of 500 function calls. Numerical problems are also indicated by the large condition number of the Hessian.

In Example 6.1 a zero initial parameter estimate was used. One would like to use an initial estimate obtained by a fast method as i.e. least squares. However, this requires a model that is linear in the parameters, i.e. (2.1) with all C_i equal to 0. The problem is that our overlapping parametrizations cannot be transferred immediately to the VDE case. All existing algorithms for transfering between canonical state space and VDE models rely heavily on the fact that one selects the first n independent rows of R^r from above (cf. Section 3). If we merely select a basis then in the VDE the i-th component of y can depend on future values of y. In Example 3.1, taking $I^{(2)} = \{1,2,3,5\}$ leads to the following equations for $y(t)$:

$$y_1(t+1) = xy_1(t) + xy_2(t) + xy_1(t+1) + \dots$$

$$y_2(t+1) = xy_1(t) + xy_2(t) + xy_2(t+1) + \underline{xy_2(t+2)} + \dots \qquad \text{cf. (3.9)}$$

This dependence on future values of y will not occur if the basis consists of the first n rows of R^r. In that case we can estimate an LS model in VDE form and then use the algorithm of i.e. Shrikhande et al. [1980] to transform it to state space form. If the best basis is unknown one would start with the generic basis: the first n rows of R^r. Even if the wanted initial basis does not consist of the first n rows one could estimate an LS model using the generic basis and then transform it to the desired one. The transformation matrix to the desired parametrization will in general not be singular. This will lead to a saving in CPU time in the combined algorithm, which for some cases may be substantial.

Up until now we have not included the input signal since it does not add any conceptual difficulties in the parametrization problem. However, it may greatly influence the conditioning of the parametrization. All publications on experiment design reflect this fact. The MSS algorithm will only detect a bad conditioning due to the model coming close to a hyperplane in the parameter space where the system cannot be represented in the current parametrization. This is illustrated in the following example where the combined algorithm is applied to real data.

EXAMPLE 6.2. The process identified is the destillation column used by Gauthier and Landau [1978]. The process contains two local PID loops. The in- and output signals of the model are:

y_1 the bottom column temperature

y_2 the head column temperature

u_1 reference temperature for y_1

u_2 reference temperature for y_2

u_3 liquid feed flow rate

u_4 vapour feed flow rate

The identifications were done on 864 datapoints. The same data were used as in Gauthier and Landau [1978]. For more details on the process see their article. There the identified model in state space form is:

$$F = \begin{pmatrix} 0 & -0.612 & 0 & 0.015 \\ 1 & 1.525 & 0 & -0.015 \\ 0 & -0.21 & 0 & -0.9 \\ 0 & 0.211 & 1 & 1.877 \end{pmatrix} \qquad G = \begin{pmatrix} 0.05 & 0 & 0.37 & 0.04 \\ 0.035 & 0 & -0.39 & -0.02 \\ 0.025 & 0.01 & 0.05 & -1.07 \\ 0.01 & 0.01 & -0.05 & 1.13 \end{pmatrix}$$

$$H = \begin{pmatrix} 0 & 1 & 0 & 0 \\ 0 & 0 & 0 & 1 \end{pmatrix}$$

Since Gauthier and Landau used an output error method no noise model is available. This model is representable in all of the structures $O^{(1)} = \{2,2\}$ $O^{(2)} = \{1,3\}$ and $O^{(3)} = \{3,1\}$. In Gauthier and Landau the structure was selected a priori by choosing the first four independent rows of R^r from above. They selected rows 1,2,3 and 4 corresponding to our model structure $O^{(1)}$. Transforming to $O^{(1)}$ gives:

$$F^{(1)} = \begin{pmatrix} 0 & 1 & 0 & 0 \\ -0.612 & 1.525 & 0.015 & -0.015 \\ 0 & 0 & 0 & 1 \\ -0.210 & 0.211 & -0.9 & 1.877 \end{pmatrix}$$

$$G^{(1)} = \begin{pmatrix} 0.035 & 0.0 & -0.39 & -0.02 \\ 0.1032 & -1.500\ 10^{-4} & -0.224 & -7.45\ 10^{-3} \\ 0.01 & 0.01 & -0.05 & 1.13 \\ 5.116\ 10^{-2} & 2.877\ 10^{-2} & -0.1261 & 1.047 \end{pmatrix}$$

$$H^{(1)} = \begin{pmatrix} 1 & 0 & 0 & 0 \\ 0 & 0 & 1 & 0 \end{pmatrix}$$

The process was identified with a zero initial parameter estimate and with both $O^{(1)}$, $O^{(2)}$ and $O^{(3)}$ as the initial model structure. The results are summarized in Tables 6.3-6.4. In the $O^{(1)}$ case the parametrization changed first to $O^{(2)}$ and then back to $O^{(1)}$. In the $O^{(2)}$ case the F matrix was unstable from iteration 23 to iteration 478! Consequently the MSS algorithm was not called during this time. This could have been avoided if the MSS algorithm was modified according to Remark 4 in Section 5. In the $O^{(3)}$ case the first parameter change was to $O^{(2)}$. The other 19 changes were back and forward between $O^{(2)}$ and $O^{(1)}$. In all three cases the algorithm converged to the same point in $O^{(1)}$.

The approximate Hessian at the minimum is ill conditioned. The number of iterations is also quite large. The ill conditioning is due to the G parameters and the relatively low frequency inputs u_3 and u_4. There are some parameters in the model which could be set to zero. The head temperature (u_2, y_2) does not noticeably influence the bottom temperature (y_1). This means that in the $O^{(1)}$ model $f_{12,1}, f_{12,2}, g_{21}$ and g_{22} can be put to zero. The spectrum for u_3 and u_4 are not or at least very poorly persistently exciting. This makes V'' ill conditioned even in the correct parametrization $O^{(1)}$.

Using an initial LS estimate reduces the computing time considerably. The results for the $O^{(1)}$ case are given in Tables 6.5-6.6. The number of function calls drops to about one third of the $O^{(1)}$ case with zero initial estimate! No parametrization changes occur. Transforming the initial LS model to $O^{(2)}$ and $O^{(3)}$ and then applying the combined algorithm gives the same results as the $O^{(1)}$ case, because the MSS algorithm changes to $O^{(1)}$ during the first iteration.

We will now conclude this section with a series of remarks commenting on both examples.

REMARK 1. No effort was made to trim the algorithm since we only wanted to show the feasibility of our approach: to estimate the parameters and the model structure at the same time by using overlapping parametrizations. For example, the approximative Hessian was not changed when a parametrization change occurred. Also quite a few (typically 10 to 20) function and gradient calls are made at the minimum to ensure that the found point is a minimum and not a saddle point.

REMARK 2. The termination criterion (gradient norm less than 10^{-6}) is quite severe, especially in Example 6.2 since there the number of estimated parameters was 32. The number of iterations could have been decreased by using a higher value on the criterion or using a different one, i.e. the relative change in the loss function.

REMARK 3. The presence of an input signal influenced the conditioning of the parametrization in Example 6.2. How an input signal influences the selection of a better conditioned parametrization should be investigated.

REMARK 4. The parametrization changes in the $0^{(1)}$ and $0^{(2)}$ cases in Example 6.2 are caused by y_2 having a higher signal power than y_1. Var y_1 = 1.47 and Var y_2 = 30.61. This makes the MSS algorithm having a preference for the model structure containing predictors of y_2, i.e. $I^{(2)} = \{1,2,4,6\}$ containing $\hat{y}_2(t+1|t)$, $\hat{y}_2(t+2|t)$ and $\hat{y}_2(t+3|t)$. In fact, the complexity for the original P matrix was less than the complexity for the new one in all parametrization changes from $0^{(1)}$ to $0^{(2)}$. This could have been avoided by introducing some kind of hysteresis in the MSS algorithm, e.g. complexity P_{old}- complexity $P_{new} > f$, where f is a positive number.

REMARK 5. Using an initial LS estimate works well. Using it in Example 6.1 gives a saving of about 20% in CPU time. Since these runs did not give anything new compared with the runs without an initial LS estimate, they are not reported here.

7. CONCLUSIONS. Parametrization of multivariable systems for identification can be made in a number of ways. Here, we have suggested that overlapping parametrizations should be used, when no a priori structural information is given. This has the advantage that the parametrization can be changed with no loss of information when numerical problems are detected or expected.

We have suggested a specific procedure based on this idea. The procedure has also been tested on real and simulated data.

The emphasis of the paper has been on the model structure selection (MSS) block of this procedure. This has been found to work satisfactorily. It does not add very much to the total computing time to monitor and possibly change the parametrization in the way suggested.

ACKNOWLEDGEMENT. The data of Example 6.2 were obtained by Dr. Bern at Rhone Poulenc. We thank him as well as Professor Landau and Mrs. Vella for their help making the data available to us.

APPENDIX

$P^{(i)}$ matrix of step 1		
$L^{(1)}$	$L^{(2)}$	$L^{(3)}$
1 0 0 0 -3.007 1 0 0 -2.646 -3.404 1 0 0.669 0.333 0.782 10^{-6} 1	1 0 0 0 -3.007 1 0 0 -2.650 -3.409 1 0 -1.175 1.333 0.847 10^{-6} 1	1 0 0 0 -3.007 1 0 0 -2.647 -3.406 1 0 -4.639 -0.071 0.081 1
$D^{(1)}$	$D^{(2)}$	$D^{(3)}$
.256 .383 .342 .313 10^{-10}	.256 .382 .344 .320 10^{-10}	.256 .382 .343 .314
Complexity of $P^{(i)}$		
31.11	31.41	9.93
Selected rows		
1,2,3,6	1,2,3,6	1,2,3,6
L matrix of result		
1 0 0 0 -3.007 1 0 0 -2.646 -3.405 1 0 -4.637 -0.070 0.078 1	1 0 0 0 -3.007 1 0 0 -2.650 -3.409 1 0 -4.646 -0.045 0.085 1	No changes
D matrix of result		
0.256 0.382 0.343 0.314	0.256 0.382 0.344 0.329	No changes
Complexity of result		
9.93	9.89	No changes
F matrix after similarity transformation		
-0.333 -0.333 0.0 10^{-5} 2.833 1.333 0.0 10^{-5} 0 0 0 1 0.334 1.333 -0.25 0.0	-0.33 -0.333 0.0 10^{-5} 2.833 1.333 0.0 10^{-5} 0 0 0 1 0.336 1.336 -0.25 0.0	No changes
K matrix after similarity transformation		
-0.0398 -0.312 -0.0012 0.3843 1.224 0.1309 -0.798 -0.236 -0.215 -0.00467 1.270 -0.215	-0.0398 -0.312 -0.0112 0.3848 1.224 0.1309 -0.798 -0.236 -0.215 -0.00415 1.271 -0.215	No changes

Table 5.1. Results from the model structure selection algorithm when applied to example 5.1.

Initial structure $O^{(1)}$, final structure $O^{(1)}$ after 0 parametrization changes.	Initial structure $O^{(2)}$ final structure $O^{(1)}$ after 1 parametrization change.
$F^{(1)} = \begin{matrix} 0 & 1 & 0 \\ 0.298 & -7.902\ 10^{-2} & 0.2842 \\ 1.042 & 3.632\ 10^{-2} & -4.453\ 10^{-2} \end{matrix}$	$F^{(1)} = \begin{matrix} 0 & 1 & 0 \\ 0.2981 & -7.907\ 10^{-2} & 0.2842 \\ 1.042 & 3.647\ 10^{-2} & -4.458\ 10^{-2} \end{matrix}$
$K^{(1)} = \begin{matrix} -8.209\ 10^{-4} & 8.436\ 10^{-2} \\ 0.1672 & 0.1736 \\ 0.6407 & -4.330\ 10^{-2} \end{matrix}$	$K^{(1)} = \begin{matrix} -8.147\ 10^{-4} & 8.435\ 10^{-2} \\ 0.1671 & 0.1736 \\ 0.6407 & -4.329\ 10^{-2} \end{matrix}$

Fixed model structure $O^{(2)}$
$F^{(2)} = \begin{matrix} -31.91 & 1.718 & 30.24 \\ 0 & 1 & 0 \\ -33.58 & 1.818 & 31.77 \end{matrix}$
$K^{(2)} = \begin{matrix} 1.587\ 10^{-3} & 8.300\ 10^{-2} \\ 0.6408 & -4.348\ 10^{-2} \\ -2.920\ 10^{-2} & 9.574\ 10^{-2} \end{matrix}$

Table 6.1. Results from Example 6.1, the identified models with $O^{(1)}$ and $O^{(2)}$ as initial model structures and with a fixed $O^{(2)}$ structure.

	Initial $O^{(1)}$, final $O^{(1)}$	Initial $O^{(2)}$, final $O^{(1)}$	Fixed $O^{(2)}$
Loss function	0.970825	0.970825	0.970900
Gradient norm	$0.4522\ 10^{-4}$	$0.1201\ 10^{-4}$	0.1369
Nr. of iterations	24	43	281
Nr. of function calls	39	59	501
Nr. of gradient calls	38	58	489
Condition nr. of Hessian	15.69	16.69	$3.249\ 10^{6}$
Complexity of P	1.652	1.652	8.552
Nr. of parametrization Changes	0	1	-----
Change in iteration(s)	-----	18	-----

Table 6.2. Results from Example 6.1, some statistics from the identification runs.

Initial structure $O^{(1)}$, final structure $O^{(1)}$ after 2 changes.	$F^{(1)} = \begin{matrix} 0 & 1 & 0 & 0 \\ -0.6008 & 1.5154 & 8.0996\ 10^{-3} & -6.6999\ 10^{-3} \\ 0 & 0 & 0 & 1 \\ -0.2314 & 0.2290 & -0.8753 & 1.8553 \end{matrix}$ $G^{(1)} = \begin{matrix} 3.9541\ 10^{-2} & 7.4324\ 10^{-3} & -0.3964 & -0.3345 \\ 0.1043 & 4.5515\ 10^{-3} & -0.2249 & -0.1740 \\ 6.3868\ 10^{-3} & 7.3966\ 10^{-2} & -5.0115\ 10^{-2} & 1.7520 \\ 4.5995\ 10^{-2} & 3.1029\ 10^{-2} & -0.1332 & 1.4698 \end{matrix}$ $K^{(1)} = \begin{matrix} 1.3211 & 6.5700\ 10^{-2} \\ 1.3757 & 0.1042 \\ 0.1342 & 1.9972 \\ 0.2660 & 2.8316 \end{matrix}$
Initial structure $O^{(2)}$, final structure $O^{(1)}$ after 1 change.	$F^{(1)} = \begin{matrix} 0 & 1 & 0 & 0 \\ -0.6007 & 1.5153 & 8.1024\ 10^{-3} & -6.7029\ 10^{-3} \\ 0 & 0 & 0 & 1 \\ -0.2304 & 0.2278 & -0.8752 & 1.8553 \end{matrix}$ $G^{(1)} = \begin{matrix} 3.9572\ 10^{-2} & 7.4542\ 10^{-3} & -0.3967 & -0.3350 \\ 0.1043 & 4.5606\ 10^{-3} & -0.2250 & -0.1741 \\ 6.4859\ 10^{-3} & 7.3134\ 10^{-2} & -5.3490\ 10^{-2} & 1.7205 \\ 4.6276\ 10^{-2} & 3.0945\ 10^{-2} & -0.1359 & 1.4430 \end{matrix}$ $K^{(1)} = \begin{matrix} 1.3203 & 6.5624\ 10^{-2} \\ 1.3740 & 0.1040 \\ 0.1327 & 1.9964 \\ 0.2618 & 2.8307 \end{matrix}$
Initial structure $O^{(3)}$, final structure $O^{(1)}$ after 20 changes.	$F^{(1)} = \begin{matrix} 0 & 1 & 0 & 0 \\ -0.6008 & 1.5154 & 8.1023\ 10^{-3} & -6.7027\ 10^{-3} \\ 0 & 0 & 0 & 1 \\ -0.2315 & 0.2291 & -0.8753 & 1.8553 \end{matrix}$ $G^{(1)} = \begin{matrix} 3.9539\ 10^{-2} & 7.4316\ 10^{-3} & -0.3964 & -0.3342 \\ 0.1043 & 4.5516\ 10^{-3} & -0.2249 & -0.1739 \\ 6.3870\ 10^{-3} & 7.4018\ 10^{-2} & -4.9975\ 10^{-2} & 1.7521 \\ 4.5993\ 10^{-2} & 3.1035\ 10^{-2} & -0.1332 & 1.4699 \end{matrix}$ $K^{(1)} = \begin{matrix} 1.3210 & 6.5726\ 10^{-2} \\ 1.3756 & 0.1042 \\ 0.1344 & 1.9971 \\ 0.2665 & 2.8313 \end{matrix}$

Table 6.3. Results from Example 6.2, the identified model with $O^{(1)}$, $O^{(2)}$ and $O^{(3)}$ as initial model structure and zero initial parameters.

	Initial $O^{(1)}$, final $O^{(1)}$	Initial $O^{(2)}$, final $O^{(1)}$	Initial $O^{(3)}$, final $O^{(1)}$
Loss function	$3.8306\ 10^{-4}$	$3.8307\ 10^{-4}$	$3.8306\ 10^{-4}$
Gradient norm	$1.3748\ 10^{-7}$	$9.4388\ 10^{-7}$	$2.3215\ 10^{-7}$
Nr. of iterations	616	724	490
Nr. of function calls	651	787	535
Nr. of gradient calls	651	787	534
Condition nr. of Hessian	$7.4383\ 10^{4}$	$2.0913\ 10^{3}$	$6.3579\ 10^{4}$
Complexity of P	13.10	13.10	13.10
Nr. of parametrization changes	2	1	20
Changes in iteration(s)	175,341	549	19, 176, 181, 184, 185, 186, 187, 189, 190, 191, 192, 195, 198, 203, 208, 213, 218, 222, 228, 233

Table 6.4. Results from Example 6.2, some statistics from the identification runs with zero initial estimate.

Estimated LS model in structure $O^{(1)}$	$F^{(1)} = \begin{matrix} 0 & 1 & 0 & 0 \\ -0.5261 & 1.4316 & -5.0047\ 10^{-3} & 6.6621\ 10^{-3} \\ 0 & 0 & 0 & 1 \\ -0.1830 & 0.1757 & -0.8915 & 1.8722 \end{matrix}$ $G^{(1)} = \begin{matrix} 3.7028\ 10^{-2} & 7.0747\ 10^{-3} & -0.4243 & -0.5408 \\ 0.1100 & 3.3658\ 10^{-3} & -0.2086 & -0.2125 \\ 1.0036\ 10^{-2} & 1.0480\ 10^{-2} & -6.5932\ 10^{-2} & 1.5907 \\ 5.0005\ 10^{-2} & 3.2093\ 10^{-2} & -0.1339 & 1.3375 \end{matrix}$ $K^{(1)} = \begin{matrix} 1.4316 & 0 \\ 1.5234 & 7.4680\ 10^{-3} \\ 0 & 1.8722 \\ 0.2581 & 2.6136 \end{matrix}$
Identified $O^{(1)}$ model after 0 changes	$F^{(1)} = \begin{matrix} 0 & 1 & 0 & 0 \\ -0.6010 & 1.5157 & 8.1454\ 10^{-3} & -6.7503\ 10^{-3} \\ 0 & 0 & 0 & 1 \\ -0.2373 & 0.2356 & -0.8748 & 1.8548 \end{matrix}$ $G^{(1)} = \begin{matrix} 3.9542\ 10^{-2} & 7.4246\ 10^{-3} & -0.3961 & -0.3357 \\ 0.1043 & 4.5518\ 10^{-3} & -0.2248 & -0.1748 \\ 6.2959\ 10^{-3} & 7.3744\ 10^{-2} & -4.4342\ 10^{-2} & 1.7411 \\ 4.5406\ 10^{-2} & 3.1043\ 10^{-2} & -0.1305 & 1.4568 \end{matrix}$ $K^{(1)} = \begin{matrix} 1.3215 & 6.5753\ 10^{-2} \\ 1.3766 & 0.1043 \\ 0.1405 & 1.9956 \\ 0.2805 & 2.8283 \end{matrix}$

Table 6.5. Results from Example 6.2, the initial LS model and the identified model with the LS model as initial estimate.

	Initial $0^{(1)}$ LS	Initial $0^{(1)}$ LS, final $0^{(1)}$
Loss function	$4.1356\ 10^{-4}$	$3.8307\ 10^{-4}$
Gradient norm	$4.1279\ 10^{-4}$	$1.3864\ 10^{-6}$
Nr. of iterations	---	204
Nr. of function calls	---	220
Nr. of gradient calls	---	215
Condition nr. of Hessian	---	$2.6128\ 10^{4}$
Complexity of P	12.58	13.09
Nr. of parametrization changes	---	0

Table 6.6. Results from Example 6.2, some statistics from the identification with an initial LS estimate.

REFERENCES

H. Akaike. *Stochastic theory of minimal realization*, IEEE Trans. on Auto. Contr., Vol. AC-19, (1974), pp. 667-674.

H. Akaike. *Canonical correlation analysis of time series and the use of an information criterion.* In System Identification: Advances and Case Studies, R. K. Mehra and D. G. Lainiotis, (Eds.), Academic Press, New York, 1976, pp. 27-96.

G. J. Bierman. *Factorization Methods for Discrete Sequential Estimation.* Academic Press, New York, 1977.

Å. Björck. *Solving least square problems by Grahm-Schmidt orthogonalization,* BIT, Vol. 7, (1967), pp. 1-21.

A. Gauthier and I. D. Landau. *On the recursive identification of multi-input, multi-output systems,* Automatica, Vol. 14 (1978), pp. 609-614.

P. E. Gill and W. Murray. *Quasi-Newton methods for unconstrained optimization,* J. Inst. Math. and Appl., Vol. 9 (1972), pp. 91-108.

P. E. Gill, W. Murray, and R. A. Pitfield. *The implementation of two revised Quasi-Newton algorithms for unconstrained optimization.* National Physical Laboratory Report NAC-11.

K. Glover and J. C. Willems. *Parametrization of linear dynamical systems--canonical forms and identifiability,* IEEE Trans. Auto. Cont., Vol. AC-19 (1974), pp. 667-674.

R. Guidorzi. *Canonical structures in the identification of multivariable systems,* Automatica, Vol. 11 (1975), pp. 361-374.

R. E. Kalman, P. L. Falb and A. A. Arbib. *Topics in Mathematical System Theory,* McGraw-Hill, New York, 1969.

R. L. Kashyap and A. R. Rao. *Dynamic Stochastic Models from Empirical Data.* Academic Press, New York, 1976.

L. Ljung and J. Rissanen. *On canonical forms, parameter identifiability and the concept of complexity.* Proc. 4th IFAC Symposium on Identification and System Parameter Estimation, Tbilisi, USSR (1976), pp. 58-69.

D. G. Luenberger. *Canonical forms for linear multivariable systems,* IEEE Trans. Auto. Cont., Vol. AC-12 (1967), pp. 290-293.

J. Rissanen. *Basis of invariants and canonical forms for linear dynamic systems.* Automatica, Vol. 10 (1974), pp. 175-182.

J. Rissanen and L. Ljung. *Estimation of optimum structures and parameters for linear systems.* Proc. Symp. Advanced School on Mathematical System Theory, Udine, Italy. Lecture Notes in Economics and Mathematical Systems, Vol. 131, Springer Verlag (1976), pp. 75-91.

V. L. Shrikhande, D. P. Mital, and L. M. Ray. *On minimal canonical realization from input-output data sequences.* IEEE Trans. Auto. Contr., Vol. AC-25 (1980), pp. 309-312.

E. Tse and H. L. Weinert. *Structure determination and parameter identification for multivariable stochastic linear systems.* IEEE Trans. on Auto. Contr., Vol. AC-20 (1975), pp. 603-613.

DEPARTMENT OF ELECTRICAL ENGINEERING
LINKÖPING UNIVERSITY
S-581 83 LINKÖPING
SWEDEN

REPRESENTATIONS OF NONLINEAR SYSTEMS: MINIMALITY, OBSERVABILITY AND CONTROLLABILITY

Arjan van der Schaft

ABSTRACT. Several representations of smooth nonlinear systems with external variables are discussed and a review is given of some results on minimality, observability and controllability in this framework.

0. INTRODUCTION. It is argued in [11] (see also these proceedings) that in the modelling of dynamical systems with external variables it is not necessary to distinguish a priori between inputs and outputs. Actually, at first instance it might be much more natural not to do this. Think for instance of electrical networks, and some physical and economical systems, where it is not immediately clear which of the external variables are the causes (inputs) and which are the effects (outputs). In [11] it is proven that for linear systems we can afterwards always split the external variables into an input part and an output part, thereby indicating that for linear systems this set-up is more general on the level of definition, but in fact *not* on the level of the actual amount of systems which can be described.

However, the situation for *nonlinear* systems (see [11, §6]) is totally different. In the first place it is possible that there always remains "hidden inputs," i.e. inputs which we cannot identify with some of the external variables. In the second place a separation between inputs and outputs can be performed a priori only *locally*. In other words, it might be possible that a certain external variable should be regarded as an input in a certain domain of working of the system, and as an output in another domain. Also, contrary to the linear case, minimal nonlinear systems with a same external behavior need not be isomorphic (in [11] this is argued for systems without differential geometric structure).

Therefore we might say that especially for nonlinear systems the set-up of [11] gives a host of interesting questions, which are sometimes also illuminating for the "normal" input-output case. The study of this set-up is also motivated by the fact that already a large class of physical systems can be naturally described in it (see [8]). The purpose of this paper is to give a review of the already obtained results. The content is mainly based on [7], to which we also refer for proofs. In §1 the general set-up and the possible representations for nonlinear systems will be discussed. In §2 minimality will be defined in a differential geometric fashion. Also the connections with the concept of controlled invariance will be given (here we will make use of [6]). It is shown how a kind of observability can be deduced from minimality, and a kind of controllabilily can be defined in a similar way. In §3 the theory will be applied to Hamiltonian and gradient systems.

1. REPRESENTATIONS OF NONLINEAR SYSTEMS. As already argued in [11], see also [1,2], a smooth nonlinear system can be defined by the following commutative diagram

$$\begin{array}{ccc} B & \xrightarrow{\quad f \quad} & TM \times W \\ & {\scriptstyle\pi}\searrow \qquad \swarrow{\scriptstyle\pi_M} & \\ & M & \end{array} \qquad (1.1)$$

Here the manifold M denotes the state space, the manifold W the space of external variables, and B is a fiber bundle about M with projection π. TM is the tangent bundle of M (i.e. positions and velocities) and π_M is the natural projection of TM onto M. Finally f is a smooth map. We will denote (1.1) by $\Sigma(M,W,B,f)$, or shortly Σ.

In local coordinates $x = (x_{11},\ldots,x_n)$ for M, $(x,v) = (x_1,\ldots,x_n, v_1,\ldots,v_m)$ for B, $w = (w_1,\ldots,w_r)$ for W. This simply gives

$$\begin{aligned} \dot{x} &= g(x,v) \\ w &= h(x,v) \end{aligned} \qquad (1.2)$$

where we have split f as $(g,h) : B \to TM \times W$, with $g : B \to TM$ and $h : B \to W$. Here v, the coordinates for the fibers of B, are seen as "input" explaining the possible evolutions of the state $(\dot{x} = g(x,v))$ and the possible values of the external variables $(w = h(x,v))$, depending on the state x which is realized at a certain moment. The usual input-output framework is recovered by taking B trivial, i.e. $B = M \times U$, with U an input manifold, and by taking $h(x,v) = (\tilde{h}(x,v),v)$. In this case we have taken $w = (w_1,w_2)$, with w_2 the inputs (which now can be identified with the coordinates of U) and w_1 the outputs. Therefore the definition above departs in two aspects from the usual one: 1) instead of $M \times U$ we take a fiber bundle B above M, 2) we do not split $W = Y \times U$, with Y an output manifold.

Henceforth we will concentrate on aspect 2), but we will say a few words about 1) (for a more elaborate discussion see [1,2], from which we also borrow the following example). Consider a particle moving on the surface of a sphere, which can be controlled in a linear way everywhere in every direction. If we would take $B = M \times U$ (in this case $S^2 \times \mathbb{R}^2$) we obtain locally

$$\dot{x} = g(x,u_1,u_2) = u_1X_1(x) + u_2X_2(x) \ , \qquad x \in S^2$$

where X_1 and X_2 are *globally* defined vector fields on S^2. However, vector-fields on S^2 always have at least one singularity, say $X_1(x_1) = X_2(x_2) = 0$. Therefore on x_1 and x_2 we can only control in *one* direction. The problem is solved by taking B a fiber bundle (non-trivial) isomorphic to TS^2 (for instance T^*S^2, because in many cases forces can be seen as cotangent variables).

We will now show how by successively adding assumptions to the general framework (1.1) and (1.2) we can distinguish between several representations, finally ending with the usual input-output representation.

I. Assume $h : B \to W$ is an immersion restricted to the fibers of B, in other words $\left(\frac{\partial h}{\partial v}\right)$ is injective. Then locally we can define coordinates $= (y,u)$ for W in which coordinatization $h(x,v) = (\tilde{h}(x,v),v)$. Hence we can identify v and u and write (1.2) as

$$\begin{aligned} \dot{x} &= g(x,u) \\ y &= \tilde{h}(x,u) \ , \qquad w = (y,u) \ . \end{aligned} \tag{1.3}$$

II. Suppose moreover that Δ_0 + ker dh is an involutive distribution on B (Δ_0 is the vertical tangent space of B, $\Delta_0(x,v) := \{X \in T_{(x,v)}B \,|\, \pi_*X = 0\}$). Then $h_*\Delta_0$ is an involutive distribution on W and we can find coordinates $w = (y,u)$ for W such that $h(x,v) = (\tilde{h}(x),v)$, or (after identifying $v = u$)

$$\begin{aligned} \dot{x} &= g(x,u) \\ y &= \tilde{h}(x) \ . \end{aligned} \tag{1.4}$$

III. Now we will assume that there is a globally defined output manifold Y, and that W is a fiber bundle above Y. Moreover $h : B \to W$ is a bundle morphism. Then locally around every point $\bar{x} \in M$, $\bar{y} \in Y$

$$\begin{aligned} \dot{x} &= g(x,u) \\ y &= \tilde{h}(x) \ . \end{aligned} \tag{1.5}$$

IV. Take the same assumptions as in III, but assume that B and W are trivial bundles, i.e. $B = M \times U$, $W = Y \times U$. Then we obtain

$$\begin{aligned} \dot{x} &= g(x,u) \\ y &= \tilde{h}(x) \end{aligned} \tag{1.6}$$

where for every constant u, $g(\cdot,u)$ is a globally defined vector field on M.

REMARKS. If $\left(\frac{\partial h}{\partial v}\right)$ is not injective we are in the situation of hidden inputs. Case III should be regarded as the normal setting for nonlinear control systems, because in this case outputs are intrinsically defined and the set of possible inputs is determined by the output (because W is a bundle above Y). This is actually also the case described in [1].

2. MINIMALITY, OBSERVABILITY AND CONTROLLABILITY. We define minimality in the following way

DEFINITION 2.1. *Let* $\Sigma(M,W,B,f)$ *and* $\Sigma'(M',W,B',f')$ *be two smooth systems. Then* $\Sigma' \leq \Sigma$ *if there exist surjective submersions* $\Phi : B \to B'$, $\phi : M \to M'$ *such that the diagram*

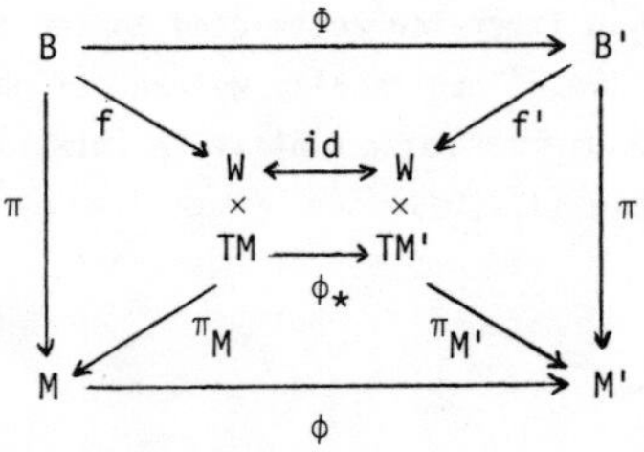

(2.1)

commutes. Σ *is called equivalent to* Σ' *(denoted* $\Sigma \sim \Sigma'$*) if* Φ *and* ϕ *are diffeomorphisms. We call* Σ minimal *if* $\Sigma' \leq \Sigma \Rightarrow \Sigma' \sim \Sigma$.

Of course, this definition formalizes the idea that we call a system minimal if we cannot make the state space or the input space smaller while retaining the same external behavior. From a differential geometric point of view it is very natural to investigate what this definition amounts to locally. Indeed, because Φ and ϕ in (2.1) are submersions we obtain that

$$E := \{X \in TB \mid \Phi_* X = 0\} \qquad \text{and}$$

$$D := \{X \in TM \mid \phi_* X = 0\}$$

are well-defined and involutive distributions on B, resp. M. Commutativeness of (2.1) is then equivalent to

$$\begin{cases} g_* E \subset \dot{D} \\ h_* E = 0 \\ \pi_* E = D \end{cases} \tag{2.2}$$

where we have again written $f = (g,h)$, and $\dot{D}$ is an involutive distribution on TM, in local coordinates constructed out of D as follows. Let $(x_1,\ldots,x_n)$ be coordinates for M such that $D = \text{span}\, \{\frac{\partial}{\partial x_1},\ldots,\frac{\partial}{\partial x_k}\}$, $k \leq n$, then $D = \text{span}\, \{\frac{\partial}{\partial x_1},\ldots,\frac{\partial}{\partial x_k}, \frac{\partial}{\partial x_1},\ldots,\frac{\partial}{\partial x_k}\}$. For a coordinate free definition see for instance [6].

Motivated by this we state

DEFINITION 2.2. *Let* $\Sigma(M,W,B,f = (g,h))$ *be a smooth system. We call* Σ locally minimal *if the only distributions* E *and* D *on* B, *resp.* M, *such that*

$$g_*E \subset \mathring{D}\ , \quad h_*E = 0\ , \quad \pi_*E = D$$

are the zero distributions.

As we saw above, if Σ is not minimal then Σ is also not locally minimal. On the other hand when Σ is not locally minimal, then it depends on the fact if E and D can be *globally* factored out (thereby obtaining B' and M') that Σ is also not minimal. Therefore local minimality is slightly stronger than minimality.

Actually, equations (2.2) are strongly related to *controlled invariance* (also called (A,B)-invariance, see [4, 5, 6]). Recall that an involutive distribution D on M is called controlled invariant with respect to the dynamics $\dot{x} = g(x,u)$ if there exists a feedback $v = \alpha(x,u)$ such that the system after feedback $x = g(x,v)$ (with $\tilde{g}(x,\alpha(x,u)) = g(x,u)$) satisfies

$$[\tilde{g}(\cdot,v),D] \subset D$$

i.e. the vectorfield $\tilde{g}(\cdot,v)$, for every constant v, leaves the distribution D invariant. In [6] it is proven that an involutive distribution D on M is locally controlled invariant iff there exists an involutive distribution E on B such that

$$g_*E \subset \mathring{D}\ , \quad \pi_*E = D \quad . \tag{2.3}$$

Furthermore, such an E exists if and only if

$$g_*(\pi_*^{-1}(D)) \subset \mathring{D} + g_*(\Delta_0) \tag{2.4}$$

where again

$$\Delta_0 = \{X \in TB \mid \pi_*X = 0\} \quad .$$

We will now show how local minimality implies a kind of observability. First we treat case I of §1 where we obtain a local input-output representation (with feedthrough term):

$$\begin{aligned} \dot{x} &= g(x,u) \\ y &= \bar{h}(x,u) \quad . \end{aligned} \tag{2.5}$$

Then we obtain

PROPOSITION 2.3. *For a system* (2.5) *local minimality implies that the only involutive distribution* D *on* M *which satisfies*

$$[g(\ ,u),D] \subset D \qquad \text{for every } u$$

$$D \subset \ker d_x\tilde{h}(\cdot,u) \qquad \text{for every } u$$

($d_x\tilde{h}$ *means differentiating with respect to* x*) is the zero distribution. We will call* (2.5) *satisfying these conditions* locally distinguishable.

For cases II, III and IV the equations are locally given by

$$\dot{x} = g(x,u)$$
$$y = \tilde{h}(x) \quad . \tag{2.6}$$

We obtain

PROPOSITION 2.4. *For a system* (2.6) *local minimality implies that the only distribution* D *on* M *which satisfies*

$$[g(\cdot,u),D] \subset D \qquad \text{for every } u$$

$$D \subset \ker d\tilde{h}$$

is the zero distribution. In other words (see [3, 5]*)* (2.6) *is locally weakly observable.*

In fact our propositions are even stronger. Consider for instance Proposition 2.4. Notice that observability can only be defined in a *specific* input-output representation $W = (y,u)$. However, take one input-output representation (2.6). Then all feedback transformations $u \to \nu = \alpha(x,u)$ which leave the form of the representation invariant (no feedthrough term), are exactly the output feedback transformations $u \to \nu = \alpha(y,u)$. Hence, apart from diffeomorphisms in the input- and output-space, another input-output representation without feedthrough term is related to (2.6) by means of output feedback. Therefore, as can be readily checked, if one such input-output representation is locally weakly observable, also all others are. Concluding, we could say that in our framework observability is a kind of derived concept, and a consequence of local minimality in every input-output representation. Since autonomous systems (i.e. without inputs) are also included in our framework we cannot hope that (local) minimality implies some kind of controllability. However we can give the following coordinate free characterization (see [7] for some technical details).

PROPOSITION 2.5. *Let* $\Sigma(M,W,B,f = (g,h))$ *be a smooth system. Then* $\dot{x} = g(x,u)$ *is strongly accessible (introduced in* [9]*, here slightly modified) if and only if there exists no nonzero distribution* D *on* M *such that*

$$g_*(\pi_*^{-1}(D)) \subset \dot{D} \quad .$$

REMARK. Notice again the relation with controlled invariance.

Finally we want to say a few words about the fact that minimal smooth systems with the same external behavior need not be equivalent. It is implicitly stated in [11, theorem 5.1] that minimal systems are equivalent if and only if the external behavior satisfies the following property: *For every feasible external behavior* $w(\cdot) : T \to W$ *there exists only* one *state evolution* $x(\cdot) : T \to M$ *such that* $(x(\cdot), w(\cdot))$ *belongs to the internal behavior of the system.*

Suppose now that we have an input-output representation, hence $w(t) = (y(t), u(t))$. Then this property simply says that when we apply the same input function to the system in state x_1 and the system in state x_2, and the resulting output functions in both cases are equal, then x_1 has to be equal to x_2. In other words, every input function distinguishes between two different states, or in the terminology of [10], every input is universal. Furthermore, we notice that this property can also be directly related to the concept of degenerate controlled invariance (see [6]).

3. HAMILTONIAN AND GRADIENT SYSTEMS. In [8] (see also [1, 2]) a coordinate free definition of a Hamiltonian and gradient system was given for a system $\Sigma(M,W,B,f)$

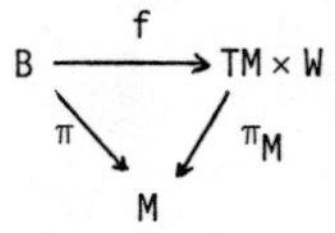

The essential feature is that we provide the state space M and the external space W with a geometric structure and ask that the submanifold $f(B) \subset TM \times W$ satisfies some properties in terms of these geometric structures. In the Hamiltonian case M and W are both equipped with a symplectic form, while for gradient systems M has a Riemannian metric on it and W again a symplectic form. In local coordinates we obtain from these coordinate free definitions the following descriptions.

Hamiltonian case:

$$\dot{q}_j = \frac{\partial H}{\partial p_j}(q,p,u)$$

$$\dot{p}_j = -\frac{\partial H}{\partial q_j}(q,p,u)$$

$$y_j = c_j \frac{\partial H}{\partial u_j}(q,p,u) \qquad c_j = \pm 1$$

with (q,p) coordinates for M, and (y,u) for W.

Gradient case:

$$\sum_i g_{ji} x_i = \frac{\partial V}{\partial x_j}(x,u)$$

$$y_j = c_j \frac{\partial V}{\partial u_j}(x,u) \qquad c_j = \pm 1$$

with x coordinates for M, and (y,u) coordinates for W. The functions $g_{ji}(x)$ form the Riemannian metric.

When the input enters linearly we obtain the following specialization (with C_j denoting the output functions)

Hamiltonian:

$$\dot{q}_j = \frac{\partial H}{\partial p_j}(q,p) + \sum_i u_i \frac{\partial C_i}{\partial p_j}(q,p)$$

$$\dot{p}_j = \frac{\partial H}{\partial q_j}(q,p) - \sum_i u_i \frac{\partial C_i}{\partial q_j}(q,p)$$

$$y_j = C_j\,(q,p)$$

Gradient:

$$\sum_k g_{jk}\dot{x}_k = \frac{\partial V}{\partial x_j}(x) + \sum_i u_i \frac{\partial C_i}{\partial x_j}(x)$$

$$y_j = C_j(x) \quad .$$

In [7] it is proven that in the *Hamiltonian* case local distinguishability implies strong accessibility and vice versa. Therefore locally minimal Hamiltonian systems are necessarily strongly accessible. However for gradient systems this is not true. In [7] a counterexample is given. Hence, despite the similarity in structure, this suggests that from a system theoretic point of view Hamiltonian and gradient systems are rather different (notice that for linear gradient systems observability *does* imply controllability, and the other way around).

REFERENCES

1. R. W. Brockett, *Global descriptions of nonlinear control problems, vector bundles and nonlinear control theory*, Notes for a CBMS Conference.

2. R. W. Brockett, *Control theory and analytical mechanics*, pp. 1-46, Geometric Control Theory (C. Martin, R. Hermann, eds.), Vol. VII, Lie Groups: History, Frontiers and Applications, Math Sci Press, 1977.

3. R. Hermann and A. J. Krener, *Nonlinear controllability and observability*, IEEE Trans. Aut. Contr., AC-22 (1977), pp. 728-740.

4. R. W. Hirschorn, (A,B)-*Invariant distributions and disturbance decoupling of nonlinear systems,* SIAM J. Cont. & Opt. 19 (1981), pp. 1-19.

5. A. Isidori, A. J. Krener, C. Gori-Giorgi and S. Monaco, *Nonlinear decoupling via feedback: A differential geometric approach,* IEEE Trans. Aut. Contr., AC-26 (1981), pp. 331-346.

6. H. Nÿmeÿer and A. J. van der Schaft, *Controlled invariance for nonlinear systems,* to appear in IEEE Trans. Auto. Contr.

7. A. J. van der Schaft, *Observability and controllability for smooth nonlinear systems*, to appear in SIAM J. Cont. & Opt.

8. A. J. van der Schaft, *Hamiltonian dynamics with external forces and observations*, to appear in Math. Systems Theory.

9. H. J. Sussmann and V. Jurdjevic, *Controllability of nonlinear systems,* J. Differential Equations, 12 (1972), pp. 95-116.

10. H. J. Sussmann, *Single-input observability of continuous-time systems,* Math. Systems Theory, 12 (1979), pp. 371-393.

11. J. C. Willems, *System theoretic models for the analysis of physical systems,* Ricerche di Automatica, Special issue on Systems Theory and Physics, Vol. X, Dec. 1979, pp. 71-106.

MATHEMATICS INSTITUTE
P.O. BOX 800
9700 AV GRONINGEN
THE NETHERLANDS

STOCHASTIC INFORMATION PROCESSING IN BIOLOGY

Harold M. Hastings and Richard Pekelney

Most current research on automata as biological models goes back to von Neumann's (1966) self-reproducing automata. Earlier, McCulloch and Pitts (1943) had used simple classical automata as formal neurons, see also Turing (1936). These authors and many others used classical deterministic automata, although probabilistic automata had been introduced in communications theory (Shannon and Weaver, 1948). Von Neumann had suggested that it would be interesting to replace classical automata by probabilistic automata, but this program was not carried out. Instead, research on probabilistic automata focused on "reliable computing with unreliable components," perhaps because vacuum tube based computers were subject to frequent Poisson failures.

We proposed (Hastings and Pekelney, 1981) that biological systems should be modelled by probabilistic automata because of the intrinsically stochastic behavior of small-scale diffusion in genetics and neurochemistry. Details and justification will be published there; we focus briefly here on the role of "errors" in biology and their import for system identification.

Mutations can be considered as errors in transcribing and/or reading genetically encoded information; however, mutations are crucial to preventing natural selection from being trapped in local maxima, and to allow rapid adaptation to unpredictable environments, cf. Conrad's (1975) adaptability theory. Similar random behavior in the brain allows for non-algorithmic learning (Conrad, 1976a,b); in the Turing (1936) ascheme, non-algorithmic learning requires replacement of one Turing machine by another. In Hastings and Pekelney (1981), we used probabilistic automata to provide at least partial explanations for these phenomena, as well as the *apparent* inefficiency of the brain and genetic systems, and the current evolutionary theory (Gould and Eldredge, 1977) of *punctuated equilibria*. Finally stochastic behavior may provide adequate solutions to any NP problems posed in environmental matching.

This discussion suggests that identification of biological systems should focus on identification of stochastic, rather than deterministic systems. In fact, many simple nonlinear models, such as the discrete-time logistic growth equation $x_{t+1} = rx_t(1 - x_t)$, $r > 3.57$, yield chaotic dynamics which prevents unambiguous classification as deterministic or stochastic (cf. May, 1979).

REFERENCES

M. Conrad, *Analyzing ecosystem adaptability*, Math. Biosci, 27:213, 1975.

M. Conrad, *Complementary molecular models of learning and memory*, Biosystems 8:119, 1976a.

M. Conrad, *Molecular information structures in the brain*, J. Neuroscience Res., 2:233, 1976b.

S. J. Gould, N. Eldredge, *Punctuated equilibria: the tempo and model of evolution reconsidered*, Paleobiology, 3:115, 1977.

H. M. Hastings, R. Pekelney, *Stochastic information processing in biological systems*, BioSystems (to appear), 1981.

R. May, *The structure and dynamics of ecological communities*, in Population Dynamics (R. M. Anderson, B. D. Turner, and L. R. Taylor, eds.), Blackwell, Oxford, 1979.

W. S. McCulloch, W. Pitts, *A logical calculus of the ideas imminent in nervous activity*, Bull. Math. Biophys., 5:115, 1943.

C. E. Shannon, W. Weaver, *The Mathematical Theory of Communication*, University of Illinois, Urbana, 1948.

A. M. Turing, *On computable numbers with an application to the Enscheidungsproblem*, Proc. London Math. Soc. Ser. 2, 42:230, 1936.

J. von Neumann, *Theory of Self-Reproducing Automata* (A. W. Burks, ed.), University of Illinois, Urbana, 1966.

ECOSYSTEMS MODELLING GROUP
DEPARTMENT OF MATHEMATICS
HOFSTRA UNIVERSITY
HEMPSTEAD, N.Y. 11550
U.S.A.

SYSTEM THEORETIC MODELS FOR THE ANALYSIS OF PHYSICAL SYSTEMS

J. C. Willems

ABSTRACT. In this paper we will introduce a new framework for the study of dynamical systems. This approach differs in an essential way from the usual input/state/output point of view in that the definitions are made without distinguishing a priori between causes (inputs) and effects (outputs). The input/output framework thus appears as a particular type of system representation. It is argued that this framework fits much better the need for providing a satisfactory conceptual basis for the modeling of physical systems in terms of the language of mathematical system theory. Various qualitative system properties (as linearity, time-invariance, etc.) are introduced from this point of view. Next we give some results regarding the state space representation of such systems. The most striking element in this context is that, contrary to what can be obtained in the classical approach, not all irreducible state space representations need be equal modulo a bijection on the state space. Several related system representation problems are subsequently treated. It is then shown how one may define dissipative systems in this context. These ideas are finally applied to thermodynamic systems. As a somewhat amusing application we also show how Newton's equations for planetary motion may be formally derived as a state space realization of Kepler's laws.

MATHEMATICS INSTITUTE
GRONINGEN
THE NETHERLANDS